Test Yourself

Introduction to Biology

Brian Mooney, M.S.
Aiken Technical College
Aiken, SC

Contributing Editors

Gloria Dyer, M.S.
Fairleigh Dickenson University
Morristown, NJ

Leane E. Roffey, Ph.D.
San Antonio, TX

Steven P. Lewis, Ph.D.
Lamar University—Beaumont
Beaumont, TX

a division of NTC Publishing Group
Lincolnwood, Illinois

Library of Congress Cataloging-in-Publication Data
is available from the Library of Congress.

A *Test Yourself Books, Inc.* Project

Published by NTC Publishing Group

Manufactured in the United States of America.

6 7 8 9 VL 0 9 8 7 6 5 4 3 2 1

Contents

Preface

Test Yourself in Biology is designed to help students in introductory biology courses. If you want higher grades, want to take tests with more confidence in your ability to do well, or want to use your study time more effectively and efficiently, this book will help you. Test-taking is a skill, and like any skill, benefits from regular practice. *Test Yourself in Biology* will give you the practice you need.

Imagine an athlete—an Olympic pole vaulter, for instance—preparing for a major event. She starts her preparation a few weeks before the contest by swimming, rowing, running, and horseback riding. In fact her preparation includes everything BUT pole vaulting. How would you rate her chances of success in the upcoming competition?

This may seem like a childishly obvious example, but it accurately describes the approach most students take to preparing for tests. They attend classes, read textbooks, listen to taped lectures, read their notes - in short, like the athlete above, they do everything before a test but practice taking a test.

Test Yourself in Biology is intended to address this shortcoming. In the following pages you will find extensive sets of questions that will test your knowledge and comprehension of the fundamental ideas of modern biology. While it is not possible to foresee the exact questions your instructor may use on a test, by surveying the various textbooks in wide use, it has been possible to construct questions that will cover the topics common in current college biology courses. As a result, students who do well on these practice tests can be confident they are well prepared to correctly answer most of the questions that are asked on examinations in the average introductory biology course. Since the coverage is comprehensive, this book will also be of value to students preparing for the Graduate Record Examination in biology or advanced placement or Regents tests.

The format of *Test Yourself in Biology* is straightforward. Each chapter contains a brief introduction, a question section, and an answer section. No attempt is made to provide a comprehensive review in the introduction, only a reminder of some of the topics covered in that chapter. Instead, explanations are provided whenever necessary along with the answers. This way, you receive help exactly where you need it, and don't have to plow through a lot of material on topics you already understand. This allows you to concentrate your time and energy where it will do the most good—on what you don't know.

Used properly, *Test Yourself in Biology* will help you find your weak spots before your instructor does (and before the weaknesses are reflected in a poor test grade). Below are some tips on how to use this book effectively:

1. Use the tests in *Test Yourself in Biology* twice in preparing for each test. Initially, use the tests to assess your knowledge. Quickly go through the relevant chapter and questions. Write answers on a separate piece of paper, and note any questions you are guessing on. This will allow you to diagnose and correct areas of weakness.

   Take each chapter test again a few days before the examination. This time, treat the practice test as if it were the actual test. This will give you a final assessment of your preparation and help you focus on remaining weak spots.

2. Answer all questions regardless of format. Even if the tests in your course are limited to multiple choice, you should attempt to answer all the questions presented. This is necessary to ensure comprehensive coverage of topics. It also reflects the fact that the value of practice in test-taking is independent of the form the questions take. Short answer and essay questions are always helpful in preparing for an examination of any sort.

   You should only omit questions that deal with material you are certain is not covered in your course.

3. Pay careful attention to diagrams, figures, illustrations, graphs, and tables in your textbook and class notes. No attempt has been made to present a large number of these in this book since the most valuable ones are those found in your textbook and notes. These are the ones most likely to show up on tests.

4. Take an active approach to learning. Passively reading your notes and textbook will not help you prepare for a test. One of the great strengths of the approach taken in Test Yourself books is to actively involve you in learning, forcing you to work with facts and ideas. You can gain most from this book by being active in all types of study. One way of doing this is to form questions yourself as you read. If the section heading for what you are reading is "the evolution of the vertebrate ear," you can make yourself an active reader by turning this into a question: "How did the vertebrate ear evolve?" "What happened during the evolution of the vertebrate ear?" Then read to answer the question.

5. Start preparing for tests days before the exam date, and make use of all resources. If your instructor has made exams for earlier semesters available, be sure to use them. Talk to your laboratory instructor or lecturer about how to prepare. While *Test Yourself in Biology* can play a central role in preparing for exams, it should form only one part of your plan for success.

6. On the actual test, be testwise. When facing multiple choice questions, be sure to read all the possible answers and cross out the wrong ones. If this type of question is difficult for you, try answering them first without looking at the answers, as if the question demanded a short answer. When you have to answer short answers or essays, begin by writing down all keywords that come to mind with regard to the questions. This list will help you give a complete, well-organized answer.

7. Condense your notes for each test. Every ten pages of class notes should be "boiled down" to no more than one or two pages of review notes. Do this by including only what you don't know in the review notes. The process of condensing will ensure a thorough review, and the resulting product will be encouragingly brief. (A set of such notes will prove especially useful in preparing for the final examination.) Over time, you will also become a more effective note taker.

Good luck!

Brian Mooney

How to Use this Book

This "Test Yourself" book is part of a unique series designed to help you improve your test scores on almost any type of examination you will face. Too often, you will study for a test—quiz, midterm, or final—and come away with a score that is lower than anticipated. Why? Because there is no way for you to really know how much you understand a topic until you've taken a test. The *purpose* of the test, after all, is to test your complete understanding of the material.

The "Test Yourself" series offers you a way to improve your scores and to actually test your knowledge at the time you use this book. Consider each chapter a diagnostic pretest in a specific topic. Answer the questions, check your answers, and then give yourself a grade. Then, and only then, will you know where your strengths and, more important, weaknesses are. Once these areas are identified, you can strategically focus your study on those topics that need additional work.

Each book in this series presents a specific subject in an organized manner, and although each "Test Yourself" chapter may not correspond to exactly the same chapter in your textbook, you should have little difficulty in locating the specific topic you are studying. Written by educators in the field, each book is designed to correspond, as much as possible, to the leading textbooks. This means that you can feel confident in using this book, and that regardless of your textbook, professor, or school, you will be much better prepared for anything you will encounter on your test.

Each chapter has four parts:

Brief Yourself. All chapters contain a brief overview of the topic that is intended to give you a more thorough understanding of the material with which you need to be familiar. Sometimes this information is presented at the beginning of the chapter, and sometimes it flows throughout the chapter, to review your understanding of various *units* within the chapter.

Test Yourself. Each chapter covers a specific topic corresponding to one that you will find in your textbook. Answer the questions, either on a separate page or directly in the book, if there is room.

Check Yourself. Check your answers. Every question is fully answered and explained. These answers will be the key to your increased understanding. If you answered the question incorrectly, read the explanations to *learn* and *understand* the material. You will note that at the end of every answer you will be referred to a specific subtopic within that chapter, so you can focus your studying and prepare more efficiently.

Grade Yourself. At the end of each chapter is a self-diagnostic key. By indicating on this form the numbers of those questions you answered incorrectly, you will have a clear picture of your weak areas.

There are no secrets to test success. Only good preparation can guarantee higher grades. By utilizing this "Test Yourself" book, you will have a better chance of improving your scores and understanding the subject more fully.

Introduction to Science and Biology

Brief Yourself

Biology is the scientific study of life, its characteristics and diverse forms, and how life has changed during the earth's long history.

Characteristics of Life

Biology rests on the chemistry and physics that make life possible. Atoms and molecules make up the non-living as well as the living world, but the living world is characterized by highly organized, complex life forms that respond to their environment, grow and take in energy and materials from the environment in order to stay alive, keep a relatively constant internal environment that is favorable to life (homeostasis), reproduce, and over time change to become better adapted to their environment by a process called evolution. The sum total of the synthetic and degradative chemical reactions carried out by a cell or organism in order to live is called metabolism.

Scientific Method and Experimentation

Scientists make statements about the world called hypotheses. While it is possible to reject a hypothesis after a single experiment, no single trial can prove a hypothesis correct. A successful series of experiments can only show that a hypothesis is more and more likely to be true.

Five Kingdoms and Scientific Naming

In biology, a very specific naming system is used to identify organisms and indicate relationships. All living things are members of one of five kingdoms (Prokarotae or Monera, Fungi, Protista or Protoctista, Plantae, Animalia). The classification for human beings is (from largest to smallest group) kingdom: Animalia; phylum: Chordata; class: Mammalia; order: Primates; family: Hominidae; genus: Homo; species: sapiens.

Unity in Diversity

The great diversity of the living world is readily appreciated, but the underlying unity that joins all life is harder to see. Organisms share some profound similarities, due to their having descended from a common ancestor. These similarities include a genetic code of DNA, and the cell as the basic unit of organization. (Some viruses have a genetic code of RNA, but viruses are parasites that only reproduce and "live" when inside the host cell, and are thus distinct from cellular life.)

Test Yourself

1. After a hypothesis has been repeatedly confirmed, and becomes generally accepted by the scientific community, it is termed a ___________.

2. Mushrooms, bread yeast, and the mold on the bathroom shower curtain are examples of organisms in the kingdom ___________.

3. Organisms that have a true nucleus, are generally free living single cells (such as amoebae or paramecium), belong to the kingdom _________.

4. Monerans or prokaryotes include ___________.

   a. viruses

   b. bacteria

   c. algae

   d. protozoa

5. Mosses and ferns are examples of what kingdom?

6. In a hierarchy of biological organization, cells with similar, specialized functions, are grouped into ________.

   a. organs

   b. macromolecules

   c. communities

   d. tissues

7. In a hierarchy of biological organization, organisms of one species which interbreed at one locality are termed a _______________.

   a. population

   b. community

   c. ecosystem

   d. biome

8. The combination of organisms and the physical environment is termed a(n) _________________.

   a. community

   b. population

   c. biosphere

   d. ecosystem

9. In a trial of a medicine to determine its effectiveness, the people receiving the medication form the _____________ group; other people who receive an identical pill which does not contain the medicine are the ___________ group.

10. In science, a possible but untested explanation is called a(n) ___________.

11. If someone says, "science can't prove that aliens haven't landed on the earth," the statement may or not be true, but is it a scientific statement? Why or why not? Please explain your answer.

12. In the scientific system of naming, the domestic dog is *Canis domestica.* In this example the species is __________ and the genus is _________.

13. Crayfish and birds are members of the kingdom _____________.

14. ____________ is the kingdom which contains living things that lack a true membrane-bound nucleus and membrane-bound organelles, and are generally single cells.

15. What steps make up the approach to learning about the world that is often called the "scientific method"?

16. List the following classification categories in correct order, from the largest and most inclusive down to the species level:

    class

    phylum (called a "division" in plants)

    species

    order

    kingdom

    genus

17. An experiment is designed to determine if a particular chemical, the artificial sweetener saccharine, increases the rate of urinary bladder cancer in laboratory rats. Identify each group.

    (i) Rats are given a normal "rat chow" diet and water to which saccharine has been added.

    (ii) Rats are given a unique diet, different from that given to other rats, and saccharine is added to their water.

    (iii) Rats are given a normal "rat chow diet and plain water.

    (iv) Rats are given a normal "rat chow" diet and water with sugar added, but no saccharine.

    (v) Mice are given a "rat chow" diet and water to which saccharine has been added.

    a. The control group is ________.
    b. The experimental group is ________.
    c. The experimental variable is ____________.

18. When a plant turns and grows in the direction of a nearby window, it is demonstrating which characteristic of living things?

19. Bacterial cells fall on a warm potato salad. After a few hours thousands of bacteria are present. This is an example of what characteristic of life?

20. As a child, you eat a hamburger and some of the protein ends up in your muscles. What two characteristics of life are demonstrated?

21. One of the most important criteria for accepting observations or results of an experiment in science is that they should be ____________.

    a. written in a book
    b. reproducible
    c. unexpected
    d. expected

22. One unifying characteristic of organisms is that genetic information (heredity) is contained in a _________ code for all living organisms.

    a. DNA
    b. protein
    c. computer
    d. atomic

Questions 23–27. True or false: If false, change the statement so that it is correct.

23. Homeostasis is a characteristic of organisms, and refers to their maintaining a relatively unstable internal environment.

24. The basic unit of life is the cell.

25. Living things are complex, and maintain their complexity by giving energy to the surrounding environment.

26. Organisms reproduce, that is, make copies of themselves including their genes.

27. Descent from a common ancestor is the underlying reason for the unity found among living things.

28. If an organism is multicellular, doesn't photosynthesize, has an organized nucleus, and secretes digestive enzymes, and absorbs the broken-down food back into the cell, the organism is a member of the _________ kingdom.

29. Organisms that are able to obtain energy from the sun and store it as chemical energy are called ____________; all other organisms are dependent upon this first type and are either ____________ or ____________.
    Possible choices for the blanks: decomposers, fungi, animals, plants, producers, protists, consumers.

30. A biologist watches baby gorillas interact with their mothers in the wild. Another biologist searches the leaves of a maple tree for insects. A third uses an instrument and a chemical test to measure the level of an enzyme in the blood of people with diabetes. All these scientists are performing what sort of scientific activity?

a. forming hypotheses

b. using instruments

c. making observations

d. forming theories

31. List, in order, the hierarchy of the organization of life, starting with atoms and ending with the biosphere.

32. People often say things like, "That pile of junk mail on my desk grows every day." But isn't growth a property of life? Why isn't the stack of mail alive?

33. What characteristic of life can explain the great diversity of organisms?

34. Animals use the simple end products of digestion to produce more complex compounds needed to get bigger and heavier. This best illustrates the life process of

a. ingestion

b. responsiveness or sensitivity

c. transport

d. metabolism and growth

35. Organisms that are very similar in structure and in the manner in which they perform life functions, but are not capable of interbreeding in nature, could be classified in the same

a. genus, but different species

b. species, but different phyla

c. phylum, but different kingdoms

d. genus, but different kingdoms

36. Which set of classification terms is shown in order of increasing numbers of organisms per group?

a. phylum, species, genus, kingdom

b. kingdom, phylum, genus, species

c. species, genus, phylum, kingdom

d. kingdom, genus, species, phylum

37. The process by which the internal environment of the cell or organism is maintained in a form favorable to life is called __________.

38. Give one example of each: a producer, a consumer, and a decomposer.

39. The process by which organisms change over generations in response to demands of the environment is called ______________.

40. Organisms that live on dead organic materials are known as:

a. parasites.

b. decomposers.

c. herbivores.

d. carnivores.

41. List the five kingdoms and give an example of each.

42. ____________ reasoning involves using a series of observations to form an hypothesis that links or explains the observations. ___________ reasoning begins with a preconception—an hypothesis—which is tested by making predictions based on that preconception.

(Choices: inductive, deductive.)

Match the kingdom named in questions 43–47 with the letter for the statement listing typical characteristics for members of that kingdom.

43. Monera or Prokaryotae	a. Eukaryotic, has a cell wall, does not photosynthesize, single or multicellular, decomposers or consumers.
44. Protista or Protoctista	b. Eukaryotic, no cell wall, does not photosynthesize, multicellular, consumers, often motile.
45. Fungi	c. Eukaryotic, has cell wall, photosynthesizes, predominantly multicellular, producers.
46. Plantae	d. Eukaryotic, some photosynthesize, generally single cells, producers or consumers, frequently motile.

47. Animalia — e. Prokaryotic, no membrane-bound structures in cells, some photosynthesize but don t have chloroplasts, producers, decomposers, or consumers.

48. Define the term "metabolism."

Check Yourself

1. Theory **(Scientific method and experimentation)**

2. Fungi. Fungi are single- or multicellular, non-motile eukaryotic decomposers or consumers such as yeasts, bread mold and mushrooms. **(Five kingdoms)**

3. Protoctista or Protista Protoctista (also called Protista) are also single-celled (unicellular) but posses both membrane-bound organelles such as mitochondria and a true nucleus enveloped by a membrane. Cells with this type of nucleus are termed eukaryotic and include all organisms in the five kingdoms but the Prokaryotae. Protoctists include algae and diatoms. Remember to learn the name preferred by your instructor. **(Five kingdoms)**

4. b. bacteria. Bacteria Prokaryotae (also called Monera) are single-celled organisms such as bacteria which do not have membrane-bound cell organelles or a true nucleus. **(Five kingdoms)**

5. Plantae Plants (Plantae) are multicellular eukaryotes which are mainly producers which photosynthesize food. Almost anything green in nature is a member of this kingdom, although not all plants are green. **(Five kingdoms)**

6. d. tissues **(Hierarchy of organization)**

7. a. population **(Hierarchy of organization)**

8. d. ecosystem **(Hierarchy of organization)**

9. experimental, control **(Scientific method and experimentation)**

10. Hypothesis **(Scientific method and experimentation)**

11. No, it is not a scientific statement: (1) it can't be tested and (2) if it is a false statement it can't be shown to be false. According to accepted criteria (first stated by the philosopher Karl Popper), for a hypothesis to be suitable it must be (1) testable, and (2) if the hypothesis is false, it must be able to be shown by testing that it is false. **(Scientific method and experimentation)**

12. *domestica, Canis* **(Scientific naming)**

13. Animalia **(Five kingdoms)**

14. Monera or prokaryotae **(Five kingdoms)**

15. The correct answer should include steps similar to these:

    (1) Initial investigation or observations. (2) Forming a testable hypothesis. (3) Experiment. (4) Collection and analysis of experimental data. (5) Interpretation of results. (6) Refine, accept or reject hypothesis. (7) Further investigation. Here the cycle begins again. **(Scientific method and experimentation)**

16. Kingdom, phylum (division in plants), class, order, genus, species. **(Five kingdoms)**

17. The control group is (iii), the experimental group is (i), and the experimental variable is saccharine. Group (ii) rats are given a unique diet, so they would not match any of the possible control groups. In group (iv), the rats are given sugar, whereas the experiment is about the effects of saccharine. In group (v), mice are used instead of rats and there is no matching control group available. **(Scientific method and experimentation)**

18. Responsiveness. The plant responds by turning toward the light. **(Characteristics of life)**

19. Reproduction. **(Characteristics of life)**

20. Metabolism and growth. **(Characteristics of life)**

21. b. Reproducible. The results from an experiment should be reproducible at other times and by other investigators. **(Scientific method and experimentation.)**

22. a. DNA Some viruses have an RNA genetic code. However, viruses must be inside a host cell to reproduce and "live" so they are usually considered separately from cellular prokaryotic and eukaryotic organisms. **(Characteristics of life)**

23. False. Replace "unstable" with "stable" or "constant" to make the statement true. **(Characteristics of life)**

24. True **(Characteristics of life)**

25. False. Replace "giving" with "acquiring." Organisms take energy from the environment in order to live. **(Characteristics of life)**

26. True **(Characteristics of life)**

27. True **(Scientific naming)**

28. Fungi **(Five kingdoms)**

29. "Producers" in the first blank; "consumers" and "decomposers" in the remaining blanks . Organisms vary in their relationship to the environment, such as how they go about obtaining materials and energy (food). Producers, such as green plants, can make food from simple inorganic compounds and sunlight. Decomposers, such as fungi, live off of waste products or dead organisms. Consumers eat producers, decomposers, or other consumers. **(Organism and environment)**

30. c. observation **(Scientific method and experimentation.)**

31. Atoms, molecules, macromolecules, cells, tissues, organs, organ systems, organisms, populations, communities, ecosystems, biosphere. **(Hierarchy of organization)**

32. The stack of mail doesn't metabolize, reproduce, exhibit homeostasis, or evolve. **(Characteristics of life)**

33. Living things have evolved from a common ancestor. **(Characteristics of life)**

34. d. metabolism and growth **(Characteristics of life)**

35. a. same genus but different species The similarity between the two organisms suggests they are closely related but not identical species. All other choices indicate a more distant relationship. **(Scientific naming)**

36. c. species, genus, phylum, kingdom **(Scientific naming)**

37. Homeostasis **(Characteristics of life)**

38. Green plant, cow or other animal, fungus. **(Organism and environment)**

39. Evolution. **(Characteristics of life)**

40. b. decomposers **(Organism and environment)**

41. Monera or Prokaryotae (examples: bacteria, cyanophyta); Protista or Protoctista (amoebae, diatoms, algae); Fungi (yeasts, mushrooms, bread molds); Plantae (ferns, mosses, flowering plants); Animalia (insects, spiders, birds). **(Five kingdoms)**

42. Inductive, deductive. Inductive reasoning means to reason from specific observations to general statements. For example, collecting data on the weather for several years and then analyzing the data to see if a general principle can be inferred. Deductive reasoning works the other way. A scientist starts with a hypothesis and makes observations or performs an experiment that will help him confirm or reject the hypothesis. In contrast to induction, deductive reasoning involves a directed inquiry that is focused on testing the hypothesis. **(Scientific method and experimentation.)**

43. e. **(Five kingdoms)**

44. d. **(Five kingdoms)**

45. a. **(Five kingdoms)**

46. c. **(Five kingdoms)**

47. b. **(Five kingdoms)**

48. Metabolism is the sum total of all the chemical reactions carried out in the process of living by a cell or organism. **(Characteristics of life)**

Grade Yourself

Circle the question numbers that you had incorrect. Then indicate the number of questions you missed. If you answered more than three questions incorrectly, you will then have to focus on that topic. (If a topic has less than three questions and you had at least one wrong, we suggest you study that topic also. Read your textbook, a review book, or ask your teacher for help.)

Subject: Introduction to Science and Biology

Topic	Question Numbers	Number Incorrect
Scientific method and experimentation	1, 9, 10, 11, 15, 17, 21, 30, 42	
Five kingdoms	2, 3, 4, 5, 13, 14, 16, 28, 41, 43, 44, 45, 46, 47	
Hierarchy of organization	6, 7, 8, 31	
Scientific naming	12, 27, 35, 36	
Characteristics of life	18, 19, 20, 22, 23, 24, 25, 26, 32, 33, 34, 37, 39, 48	
Organism and environment	29, 38, 40	

Introduction to Chemistry and the Chemistry of Life

Brief Yourself

Elements and Subatomic Particles

There are more than 108 forms of matter—the elements. Each is represented by a different chemical symbol (such as H for hydrogen, Na for sodium). Atoms of each element have a unique number of protons (positively-charged subatomic particles), and unless the atom is electrically charged (and therefore called an ion), it also has an equal number of negative charges (electrons). The number of neutral subatomic particles varies from element to element, and many elements have subgroups called isotopes, which differ only in the number of neutrons. In all cases, however, the number of protons is unique to a specific element, and referred to as its atomic number. The number of protons plus the number of neutrons is that isotope's atomic mass number.

Atoms and Molecules, Bonds

Atoms are usually combined together in groups referred to as molecules. The atoms are tied together by one of two types of chemical bonds which involve the sharing or loss and gain of electrons. Atoms are joined by a covalent bond (shared pair or pairs of electrons) or an ionic bond in which ions (atoms which have lost or gained an electron(s)) of opposite charge are mutually attracted. Covalent bonds may be single (one pair of electrons are shared), double (two pairs of electrons are shared), triple (three pairs are shared) or even quadruple (four pairs are shared), and the sharing may be unequal, lending an ionic character to the covalent bond (polar covalent bond). Carbon is a uniquely useful atom in biological molecules because of its capacity to form four covalent bonds with other elements or between carbon atoms.

Hydrogen Bonds

Hydrogen bonds are common in biological molecules, but are not true bonds; rather they are an attraction that a hydrogen atom attached to a carbon or nitrogen atom in one molecule has for an oxygen or nitrogen atom in another molecule. Hydrogen bonding is common in water and accounts for many of the special properties of water. They are also of critical importance in many proteins, stabilizing their often complex forms.

Organic Molecules

Organic molecules are carbon-based, and while they vary greatly in composition, fall into four basic groups: the carbohydrates (made up of carbon, hydrogen, and oxygen), lipids (fats and steroids), proteins (made up of amino acids), and nucleic acids (RNA and DNA). Large organic molecules are frequently made up of smaller, very similar or even identical units or monomers such as simple sugars (monosaccharides) or amino acids (in proteins). The larger molecules are termed polymers.

Test Yourself

1. The nucleus of the atom is made up of subatomic particles called __________ and __________, whereas __________ are found in the space around the nucleus.

2. Fill in the blanks with the name of the appropriate subatomic particle: __________ have a positive charge, __________ have no charge and __________ have a negative charge.

3. Circle the correct answer from each pair of choices: Anions are atoms that have (lost / gained) one or more electrons to become (positively / negatively) charged. Cations are atoms that have (lost / gained) one or more electrons to become (positively / negatively) charged.

4. All atoms or ions of an element have the same number of __________.

5. __________ __________ (two words) refers to the number of protons in an atom; __________ __________ number (two words) refers to the number of protons plus the number of neutrons.

6. Using your knowledge of isotopes, and the information given below, complete the following table. All atoms are uncharged. Hydrogen-1 is completed as an example.

isotope	number of protons	number of electrons	number of neutrons	mass number
Hydrogen-1	1	1	0	1
Hydrogen-2	1	1	___	2
Carbon-13	___	___	6	___
Carbon-14	___	___	___	14
Oxygen-16	8	___	___	___

7. A molecule of glucose has the formula $C_6H_{12}O_6$. This means each molecule has _____ carbon atoms, ____ hydrogen atoms, and ____ oxygen atoms. Written as 2 $C_6H_{12}O_6$ it represents two __________ of glucose.

8. The mass (or weight) of a molecule is calculated by __________ together the individual atomic mass numbers (amu) of the atoms making up that element. For example, if a hydrogen atom has an amu of 1 and an oxygen atom has an amu of 16, the __________ __________ for water H_2O would be ______. If the atomic mass of carbon is 12, the __________ __________ of glucose ($C_6H_{12}O_6$) would be __________.

9. Within a water molecule, the hydrogen-to-oxygen bond features an unequal sharing of a pair of bond electrons. This is an example of a __________ bond.

   a. hydrogen
   b. covalent
   c. polar covalent
   d. ionic

10. Between water molecules there is an attraction between oxygen and hydrogen atoms. This is called a __________ bond and is important in many biological molecules.

    a. hydrogen
    b. covalent

c. polar covalent

d. ionic

11. The attraction between positively-charged sodium ions and negatively-charged chloride ions in table salt is an example of a(n) __________ bond.

a. hydrogen

b. covalent

c. polar covalent

d. ionic

12. The pH scale runs from pH ___ to pH ___. On this scale, neutral is pH ____, acid is any pH value less than ______ and basic is any pH value more than ____. The pH scale is based on the concentration of __________ ions.

13. The difference in acidity in going from one unit to another on the pH scale is a factor of ______. That means a change from pH 7 to pH 4 would be a _____ increase in acidity.

a. 10 X

b. 100 X

c. 1000 X

d. 10000 X

14. ____________ bonds are only about 5 percent the strength of covalent bonds, and frequently stabilize the three dimensional structure of complex biological molecules such as proteins and DNA.

a. hydrogen

b. covalent

c. polar covalent

d. ionic

15. In the following chemical reaction for photosynthesis, label each side of the arrow as "reactants" or "products." How many molecules of water and of carbon dioxide (CO_2) are found on the left side?

light + 12 H_2O + 6 CO_2 → ($C_6H_{12}O_6$) + 6H_2O + 6O_2

16. The electrons involved in chemical bonds are found in the __________(outermost / innermost) or _____________ shell of an atom. An atom that has this shell filled with _______ is chemically unreactive.

17. Chemical reactions can be of two types with regard to energy: _________ reactions that give off energy to the surrounding environment and ___________ reactions that require energy from the environment for the reaction to occur.

18. A base is a compound which __________ the number of hydrogen ions or increases the number of ________ ions in a solution.

19. The outermost electron shell of an atom of a stable element such as neon (Ne) has _____ electrons. By acquiring, giving up or sharing electrons in chemical _______, atoms achieve this stable configuration in which the _________ shell is full.

20. Compounds which minimize pH changes in a solution (such as blood) are called __________.

21. Complete the following table:

element	number of bonds formed
Carbon	______
Oxygen	______
Hydrogen	______
Sulfur	______
Nitrogen	______
Phosphorus	______

22. The two most common elements in organic molecules, other than carbon, are _________ and _______.

23. ______________ are organic compounds which include the elements _______ and ________ in a 2:1 ratio.

24. The simplest carbohydrates are _______________, usually with a formula of $C_X(H_2O)$. Units of these are the _________ in long polymer molecules called polysaccharides.

25. Glucose is a simple sugar with the formula $C_6H_{12}O_6$. _________ of glucose which

have the same empirical formula but different structural formulas are _________ and ___________.

26. Complete this table:

disaccharide	made up of these monosaccharides
sucrose	_________ and _________
_________	glucose and glucose
_________	glucose and _________

27. Complete the following table:

polysaccharide	function	organism	where located
starch	______	______	______
______	______	plants	cell walls
chitin	______	______	______
______	storage	______	muscles, liver

28. Match the words in list A with those items in list B.

List A

a. polyunsaturated fat _____
b. fatty acid _____
c. triglyceride _____
d. saturated fat _____
e. unsaturated fat _____
f. phospholipid _____

List B

i) Glycerol backbone with two fatty acid chains and a phosphate group as the fundamental structure

ii) No double bonds between carbon atoms in a molecule

iii) Glycerol backbone linked to three fatty acid molecules

iv) One double bond between carbon atoms in a molecule

v) General formula of $CH_3(CH2)_nCH_2COOH$, where n is usually an even number

vi) Two or more double bonds between carbon atoms in a molecule

29. In which equation would X most likely represent a disaccharide?

a. amino acid + amino acid → X + water
b. fatty acid + fatty acid + fatty acid + glycerol → X + water
c. maltose + sucrose → X + water
d. glucose + glucose → X + water

30. The division of large complex molecules into smaller simpler molecules is accomplished by a process known as

a. pinocytosis
b. hydrolysis
c. absorption
d. condensation

31.

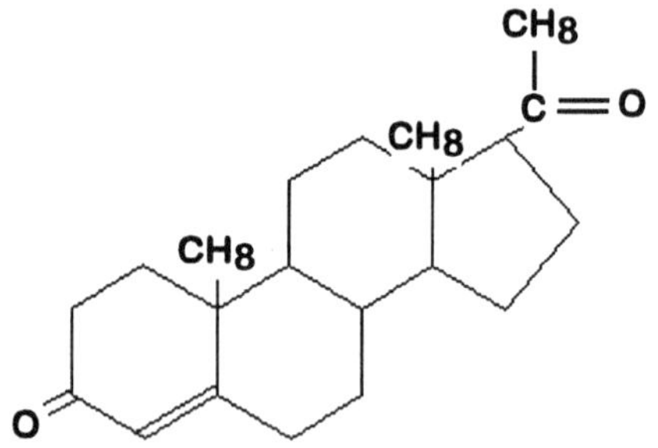

CH8
C=O
CH8
CH8
O

The formula indicated in the figure above represents an organic compound known as

a. a carbohydrate
b. a protein
c. an amino acid
d. a lipid

32.

a. What number points to a carboxylic group?

b. What number points to an amino group on this molecule?

c. What is this molecule of?

33.

a. Is this a purine or pyrimidine molecule?

b. Is this a purine or pyrimidine molecule?

34.

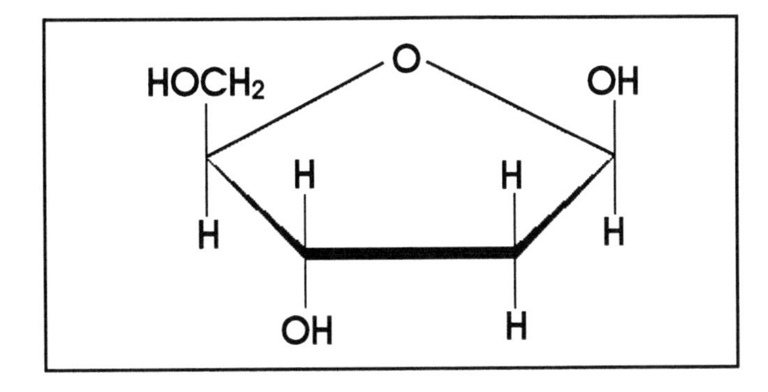

The structural formula shown here represents a molecule of ___________.

a. carbohydrate

b. lipid

c. protein

d. nucleic acid

35. Match the description in list A with the level of protein structure in list B.

List A

a. folding or conformation of the entire protein molecule in space

b. sequence of amino acids in a protein

c. protein made up of two or more polypeptide chains (subunits)

d. coil (alpha-helix), folded sheet (beta-pleated sheet), or random

List B

i) primary (1st)

ii) secondary (2nd)

iii) tertiary (3rd)

iv) quaternary (4th)

v) pentanary (5th)

36. Which one of the following elements is always found in nucleic acids but never found in amino acids?

    a. nitrogen

    b. sulfur

    c. phosphorus

    d. oxygen

37. The insulin molecule consists of two polypeptide chains, which are held together by covalent bonds formed between sulfur atoms in each chain.

    a. What is this sort of covalent bond between sulfur atoms called?

    b. Which levels of protein structure are found in the insulin molecule?

38. Steroids (sterols) have a complex structure of four rings that look like the floor plan to a house. They are examples of

    a. lipids

    b. proteins

    c. carbohydrates

    d. nucleotides

39. The element sodium has an atomic number of 11. A sodium ion has _______ protons and _______ electrons. A sodium ion (Na+) has ________ protons and ________ electrons.

40. There are _______ naturally occurring amino acids found in proteins. Each amino acid has a similar structure: an amino group, an acid group, a central carbon atom, and a(n)_________ which accounts for the chemical differences between amino acids.

41. Two monosaccharides, glucose and fructose, are chemically combined to make a disaccharide molecule. Glucose and fructose (isomers) each have a molecular weight (mass) of 180, and 2(180) = 360. But the molecular weight of sucrose is only 342. Why?

42. The nucleic acids DNA and RNA each consist of four nucleotides which differ in their base composition. The three bases found in nucleotides in both DNA and RNA are ________, _________, and _________ . _________ is only found in DNA and ________ is unique to RNA.

43. Nucleotide bases can be classified into two major groups:

    a. acids and bases

    b. ribose and deoxyribose

    c. purines and pyrimidines

    d. mRNA and tRNA

44. Each nucleotide consists of a _____________, a _________, and a _________ group.

    a. nitrogen-containing base, 5-carbon sugar, phosphate

    b. nitrogen-containing base, 6-carbon sugar, phosphate

    c. nitrogen-containing base, 5-carbon sugar, purine

    d. nitrogen-containing base, 5-carbon sugar, pyrimidine

45. Name an isomer of:

    fructose ________

    maltose ________

46. ________________ are forms of an element which are unstable. These atoms decay over time, and in doing so release radiation.

47. In redox reactions, atoms which lose electrons are termed __________; atoms which gain electrons are called ____________.

48. With the exception of hydrogen, the elements most frequently found in organisms are most stable when they posses ________ electrons in their outmost or __________ shell.

49. Water is a unique compound. Water boils at a higher temperature than would be expected from its chemical makeup. It takes a relatively large amount of energy to raise the temperature of a gram of water one degree Celsius, or to vaporize water as in boiling. Water expands when it freezes. It is an excellent solvent for many ionic and polar compounds. Water exhibits a high surface tension, so it forms drops on a waxed (hydrophobic) surface, and will support the weight of small insects

walking on the surface of a pond. All these qualities of water can be attributed to one characteristic, _______ ___________. (two words)

50. Chemical potential energy is stored in __________ ___________. (two words)

51. Hydrogen bonds can be formed between a ___________ or ___________ atom on one molecule (which possess a partial negative charge) and a hydrogen on another molecule which is covalently bonded to another ___________ or ___________ atom. The hydrogen bears a partial positive charge due to the polar nature of this covalent bond.

52. In a strand of DNA or RNA, the ___________ of one nucleotide is covalently bonded to the __________ of another nucleotide. (pick two answers)

    a. base

    b. nitrogen-containing base

    c. sugar

    d. phosphate group

53. The difference in the names DNA and RNA reflects the difference in the __________ found in the nucleotides in each nucleic acid.

54. Proteins may be modified by the cell after synthesis. Lipoproteins have ___________ combined with the protein; when a carbohydrate is attached to a protein, the resulting molecule is called a ___________. (pick two answers)

    a. glycoprotein

    b. lipoprotein

    c. carbohydrate

    d. lipid

55. ________ carbon-to-carbon bonds allow free rotation of the carbon atoms with respect to each other, while ________ bonds restrict the movement of the atoms, fixing them in place relative to one another.

56. Hydrocarbons are chains of ________ atoms with only _________ atoms attached.

57. __________ , such as bees produce, are lipids with long fatty acid chains attached to alcohols or ring-containing carbon compounds.

58. If a molecule—e.g. a protein or a nucleic acid—suffers disruption of its three-dimensional conformation and a loss of function, such as from exposure to excessive heat or extremes of pH, it is _____________. This can be reversible or irreversible.

59. Match the formulas in List B with the correct names in List A:

List A

a. phosphate group ____

b. carboxylic group ____

c. sulfhydryl group ____

d. methyl group ____

e. amino group ____

f. fatty acid ____

g. hydroxyl group ____

List B

(i) $-CH_3$

(ii) $-COOH$

(iii) $-SH$

(iv) $-NH_2$

(v) $-PO_4^{2-}$

(vi) $-OH$

(vii) $H\text{-}O\text{-}H$

(viii) $CH_3(CH_2)_{16}COOH$

Check Yourself

1. protons, neutrons, electrons **(Elements and subatomic particles)**

2. protons, neutrons, electrons **(Elements and subatomic particles)**

3. gained, negatively, lost, positively. Anions (remember as " A - N(egative)- ION) are atoms which gain one or more electrons and become negatively-charged ions (Cl^-, I^-). Cations lose one or more electrons and become positively charged since there are now more protons than electrons in the ion (e.g. Na^+, Ca^{++}). **(Elements and subatomic particles)**

4. protons **(Elements and subatomic particles)**

5. atomic number, atomic mass **(Elements and subatomic particles)**

6.

isotope	number of protons	number of electrons	number of neutrons	mass number
Hydrogen-1	1	1	0	1
Hydrogen-2	1	1	1	2
Carbon-13	6	6	7	13
Carbon-14	6	6	8	14
Oxygen-16	8	8	8	16

Explanation: (1) All atoms and ions of an element have the same atomic number that is, the same number of protons. This is a characteristic shared by all atoms of that element. (2) Atoms or ions of the same element may or may not have the same number of neutrons. Isotopes are forms of an element that have different numbers of neutrons. (3) Atoms (by definition uncharged) of an element all have the same number of electrons, equal to the number of protons. By becoming an ion, however, an atom may gain or lose one or more electrons. **(Elements and subatomic particles)**

7. 6, 12, 6, molecules **(Organic molecules)**

8. adding, molecular mass (or weight), 18, molecular mass, 180. Explanation: A molecule of glucose $C_6H_{12}O_6$ is made up of 12 hydrogen atoms (amu = 1), 6 oxygen atom (amu = 16) and 6 carbon atoms (amu = 12).

   12 H atoms times 1 amu = 12

   6 O atoms times 16 amu = 96

   6 C atoms times 12 amu = 72

   Molecular mass of glucose = 180 (cm) **(Chemical bonds)**

9. c. polar covalent **(Chemical bonds)**

10. a. hydrogen **(Chemical bonds)**

11. d. ionic **(Chemical bonds)**

12. 0, 14, 7, 7, 7, hydrogen **(pH, acids and bases)**

13. a, c. Each step on the pH scale is a factor of 10 X. Three such steps (pH 7 to 6, 6 to 5, and 5 to 4) involves three changes of 10 X each, or 10 x 10 x 10 = 1000. **(pH, acids and bases)**

14. a. hydrogen **(Chemical bonds)**

15. The reactants are on the left side of the arrow, the products are on the right side. The number (coefficient) 12 before H_2O indicates twelve water molecules, and the coefficient before the CO_2 indicates six molecules. **(Chemical reactions)**

16. outermost, valence, electrons **(Chemical reactions)**

17. endergonic, exergonic. Endothermic and exothermic may be used to describe reaction where the energy is taken in or given off as heat. **(Chemical reactions)**

18. reduces or decreases, hydroxyl (OH-) **(pH, acids and bases)**

19. eight, bonds, valence **(Elements and subatomic particles)**

20. buffers **(pH, acids and bases)**

21. carbon forms four, oxygen two, hydrogen one, sulfur two, nitrogen three, phosphorus three **(Chemical reactions)**

22. oxygen, hydrogen **(Organic molecules)**

23. carbohydrates, hydrogen, carbon. Ratio of hydrogen to carbon is the same is found water. **(Carbohydrates)**

24. monosaccharides, monomers **(Carbohydrates)**

25. isomers, fructose, galactose **(Carbohydrates)**

26. The disaccharide sucrose is made up of a molecule each of the monosaccharides glucose and fructose. Maltose is made up of two glucose molecules. The remaining disaccharide is lactose, which is made up of glucose and galactose. **(Carbohydrates)**

27. The completed table: **(Carbohydrates)**

polysaccharide	function	organism	where located
starch	storage	plants	tubers
cellulose	structural	plants	cell walls
chitin	structural	fungi; animals	cell walls; exoskeleton of arthropods
glycogen	storage	animals	muscles, liver

28. List A. **(Organic molecules)**

    a. Polyunsaturated fat _vi_

    b. Fatty acid _v_

    c. Triglyceride _iii_

    d. Saturated fat _ii_

    e. Unsaturated fat _iv_

    f. Phospholipid _i_

29. d. In a condensation (dehydration synthesis) reaction glucose + glucose will give one molecule of maltose plus a molecule of water that was removed in forming a bond between the two monosaccharides. Answer (a) is a similar reaction but the question asked for a disaccharide. **(Chemical reactions)**

30. b. Hydrolysis ("splitting with water") is a process of breaking down molecules into simpler compounds and using water to fulfill the new bonding requirements. This is the reverse of the reactions shown in answers (a), (b) and (d) of question 29 above. **(Chemical reactions)**

31. d. a lipid. This is a steroid (note how it looks a lot like the floor plan of a house), which is classed as a lipid. **(Lipids)**

32. a. 3

    b. 1

    c. amino acid **(Amino acids and proteins)**

33. a. pyrimidine (smaller structure has longer name) b. purine (larger structure has shorter name) **(Nucleotides and nucleic acids)**

34. a. This is the ring form of a monosaccharide molecule. **(Carbohydrates)**

35. In distinguishing the levels of protein organization, it often helps to form a mental image of a protein. Several are suggested below.

    a. iii. This can be pictured as the folding of a coiled telephone cord, as might be done to fit it into a pocket.

    b. i. Picture links in a chain, each link being an amino acid.

    c. iv. Not all proteins exhibit this level of organization, only those that require two or more subunits to be combined in order to form a functional molecule.

    d. ii. An alpha-helix is like a coiled telephone cord. A beta-pleated sheet resembles a sheet of paper which is folded repeatedly, accordion-fashion.

    There is no "pentanary" level. **(Amino acids and proteins)**

36. c. phosphorus **(Nucleotides and nucleic acids, Amino acids and proteins)**

37. a. disulfide bridge

    b. Insulin shows primary, secondary, tertiary levels of organization, and since it is formed of polypeptide subunits, it has quaternary structure as well. **(Amino acids and proteins)**

38. a. lipids. Steroids are classified as lipids, although they lack the characteristic fatty acid chains. Like other lipids, the non-polar steroids are insoluble in water (a polar molecule) and soluble in non-polar organic solvents as well as other lipids. **(Lipids)**

39. 11, 11, 11, 10. All sodium atoms or ions have the same number of protons, i.e., the same atomic number. A sodium atom has 11 protons and 11 electrons, making it electronically neutral. A sodium ion (Na+) has lost one electron to become a monovalent cation (an ion with a single positive charge). **(Elements and subatomic particles)**

40. 20, R-group. The R-group can stand for anything from a single additional hydrogen atom (in glycine) to chains of three carbon and one nitrogen atoms (lysine) or ring structures (phenylalanine, tyrosine). **(Amino acids and proteins)**

41. Each monosaccharide has a molecular mass of 180, and each is covalently bonded to the other in a dehydration synthesis (condensation) reaction in which a water molecule is given up in forming the new bond. The molecular mass of water is 18, and 360 - 18 = 342. **(Chemical reactions)**

42. adenine, cytosine, guanine. Thymine is found only in DNA and uracil only in RNA. **(Nucleotides and nucleic acids)**

43. c. purines and pyrimidines **(Nucleotides and nucleic acids)**

44. a. nitrogen-containing bases, 5-carbon sugar, phosphate **(Nucleotides and nucleic acids)**

45. fructose: glucose, galactose. Each sugar has the molecular formula of $C_6H_{12}O_6$ but a different structural formula.

    maltose: sucrose, lactose. Each disaccharide has a molecular formula of $C_{12}H_{22}O_{11}$ but a different structural formula. **(Carbohydrates)**

46. Radioisotopes or radioactive isotopes. **(Elements and subatomic particles)**

47. oxidized, reduced. "Redox" is a term used for such reactions since an oxidation is always coupled with a reduction. An easy way to know which atom loses and which atom gains electrons is to recall, "LEO the lion says GER." That is, L(ose) E(lectrons) O(xidized) G(ain) E(electrons) R(educed) **(Chemical reactions)**

48. eight, valence **(Elements and subatomic particles)**

49. Hydrogen bonding—an attraction between an oxygen atom with a partial negative charge on one molecule and a hydrogen atom on another water molecule bearing a partial positive charge. This attraction holds water molecules together, accounting for waters tendency to form drops and have a high surface tension. The partial charges are also very attractive to Na+ and Cl- ions, causing NaCl to spontaneously dissociate in water. Since water expands slightly on freezing, ice is less dense than water and ice forms on top of ponds, rivers and lakes, not on the bottom. **(Chemical bonds)**

50. chemical bonds. When the bonds are broken, energy is released (potential energy becomes kinetic energy). **(Chemical bonds)**

51. nitrogen, oxygen, nitrogen, oxygen **(Chemical bonds)**

52. c. sugar, and d. phosphate group. Covalent bonds hold together the nucleotides within one molecule (or one strand in double-stranded DNA) of a nucleic acid. In double-stranded DNA (double helix) hydrogen bonds hold the two chains together by spanning from a nucleotide base on one chain (A, T, C, G) to another base on the other, complementary chain of nucleotides. **(Nucleotides and nucleic acids)**

53. sugars. Deoxyribose is found in DNA, and ribose is found in RNA. **(Nucleotides and nucleic acids)**

54. d. lipid and a. gycloprotein **(Amino acids and proteins)**

55. single, double **(Chemical bonds)**

56. carbon, hydrogen **(Organic molecules)**

57. waxes **(Lipids)**

58. denatured **(Amino acids and Proteins)**

59\.

List A

a. phosphate group 5

b. carboxylic group 2

c. sulfhydryl group 3

d. methyl group 1

e. amino group 4

f. fatty acid 8

g. hydroxyl group 6

List B

(i) $-CH_3$

(ii) $-COOH$

(iii) $-SH$

(iv) $-NH_2$

(v) $-PO_4^{2-}$

(vi) $-OH$

(vii) H-O-H

(viii) $CH_3(CH_2)_{16}COOH$

(Organic molecules)

List A a. phosphate group 5

Grade Yourself

Circle the question numbers that you had incorrect. Then indicate the number of questions you missed. If you answered more than three questions incorrectly, you will then have to focus on that topic. (If a topic has less than three questions and you had at least one wrong, we suggest you study that topic also. Read your textbook, a review book, or ask your teacher for help.)

Subject: Introduction to Chemistry and the Chemistry of Life

Topic	Question Numbers	Number Incorrect
Elements and subatomic particles	1, 2, 3, 4, 5, 6, 19, 39, 46, 48	
Organic molecules	7, 22, 28, 56, 59	
Chemical bonds	8, 9, 10, 11, 14, 49, 50, 51, 55	
pH, acids and bases	12, 13, 18, 20	
Chemical reactions	15, 16, 17, 21, 29, 30, 41, 47	
Carbohydrates	23, 24, 25, 26, 27, 34, 45	
Lipids	31, 38, 57	
Amino acids and proteins	32, 35, 36, 37, 40, 54, 58	
Nucleotides and nucleic acids	33, 36, 42, 43, 44, 52, 53	

Cell Structure and Function

3

Brief Yourself

Cells and Cell Theory

The cell is the basic unit of life. This is true for single-celled organisms such as bacteria or protozoa (protists) as well as for multicellular plants, animals and fungi. The cell theory consists of this point and the proposition that all cells are derived from other cells.

Prokaryotic and Eukaryotic Cells

The living world can be clearly divided into two major groups. The prokaryotes (bacteria and cyanobacteria) are unicellular organisms that lack a true nucleus or other membrane-bound subcellular organelles. Eukaryotes (protists, fungi, plants and animals) possess a true nucleus—a completely separate, membrane-enclosed region of the cell containing the genetic material (DNA). In prokaryotic cells the DNA is confined to a nucleoid region, but without a delimiting membrane. Additionally, prokaryotes have a single loop of circular DNA, which is not a true chromosome in the sense of eukaryotic chromosomes.

A typical eukaryotic cell also contains large amounts of membranes which break the cell up into discrete but interconnected compartments. Some of the more important membrane structures in the cell include the endoplasmic reticulum (a site for synthesis and transport), the Golgi apparatus (where molecules are modified and packaged), and the plasma membrane, which controls access to and egress from the cell's interior, and the nuclear envelope (which is a double membrane structure punctured by nuclear pores).

Eukaryotic Cell Organelles

Eukaryotic cell organelles include Golgi bodies, endoplasmic reticulum, chloroplasts, mitochondria, vesicles, vacuoles, lysosomes, and ribosomes. The cytoskeleton is a protein framework that organizes the eukaryotic cell, and helps break it into compartments. The cytoskeleton also provides for cell shape and movement.

Test Yourself

1. According to the cell theory, all cells come from ___________ cells, and the cell is the smallest and most __________ unit of life.

2. Cells were first described and named in 1665 by

   a. Charles Darwin

   b. James Watson

   c. Louis Pasteur

   d. Robert Hooke

3. In 1838 and 1839, cells were proposed as the basis of organization in plant and animal tissues, respectively, by two German biologists named

   a. Smith Brothers

   b. Pasteur and Darwin

   c. Schleiden and Schwann

   d. Raven and Johnson

4. The size of cells is primarily determined by

   a. the size of the organism

   b. the limitations imposed by surface-to-volume ratio

   c. how many cells are in an organism—few and larger cells versus many and smaller cells

   d. whether the cells are from plants or animals

5. All cells have three characteristics in common:

   a. a cell or plasma membrane, cytoplasm, and genetic information in DNA

   b. a cell wall, cytoplasm, and genetic information in DNA

   c. a cell or plasma membrane, cytoplasm, and information in RNA

   d. a cell membrane, cytoplasm, and a true nucleus with genetic information in DNA

6. Eukaryotic cells are about _____ times larger than prokaryotic cells. If it is necessary to examine smaller intracellular details in either type of cell, a(n) ___________ microscope is used.

   a. 1000, light

   b. 1000, electron

   c. 10, light

   d. 10, electron

7. Contrast prokaryotic with eukaryotic cells in answering the questions below. "a" is already completed as an example.

   a. Give examples of the organisms of each type by kingdom—answer: protists (e.g., amoeba), plants (e.g., an oak), fungi (e.g., mushrooms) and animals (e.g., humans) represent the Eukaryotes, while Monerans or prokaryotes (e.g., bacteria) represent the prokaryotes.

   b. Do both types (prokaryote versus eukaryote) have cell membranes?

   c. Do both types have cell walls? (Account for each kingdom in your answer.)

   d. Do both have membrane-bound cell organelles?

   e. Chromosomes?

   f. Circular DNA, but no true chromosomes (meaning DNA-containing chromosomes complexed with histones)?

   g. DNA as genetic material?

   h. Subdivision of cells into membrane-bound compartments?

   i. What are the relative sizes of ribosomes in prokaryotes versus Eukaryotes?

8. What is the origin of chloroplasts and mitochondria according to the Endosymbiotic (symbiotic) theory? What are some points of evidence in favor of this theory?

9. What are the the three main features (structures or regions) within a mitochondrion?

10. _________ ____________ occurs in mitochondria, and results in the release of __________ and the production of _____ molecules from _______.

    a. Cellular reproduction, hydrogen, ATP, ADP

    b. Cellular respiration, energy, ATP, ADP

c. Cellular respiration, energy, ADP, ATP

d. Cellular respiration, oxygen, ATP, ADP

For questions 11–14, answer true or false. If false, change the statement to make it true.

11. Mitochondria are often called the "power plants of the cell."

12. In plants, in addition to chloroplasts there are other plastids, such as starch-containing amyloplasts.

13. All prokaryotic cells have mitochondria.

14. The liquid surrounding the grana in mitochondria is called the stroma.

15. The nuclear ___________ (one word) surrounds the nucleus and consists of _______ (one, two) lipid membranes.

16. Under the electron microscope, openings from the cytoplasm into the interior of the nucleus are seen, which are called _______ _________ (two words). At these places the two lipid bilayers meet and form holes in conjunction with ___________, which regulate the passage of materials such as proteins _____ (into, out of) the nucleus and RNA compounds ____ (into, out of) the nucleus.

17. In a drawing of a typical chloroplast, the stacks of flattened sacs or disks are called __________; each individual sac is a ____________. Like the mitochondrion and the nucleus, the chloroplast is surrounded by a __________ (single, double) membrane.

18. The liquid portion of the cell is termed _________; within the nucleus it is called ___________.

19. In Eukaryotes, DNA is located inside the __________ in conjunction with proteins forming a complex of tiny threads called __________. During cell division this material condenses into compact structures called _______________, meaning "colored body."

20. Which of the following is not true of the nucleolus?

    a. The nucleolus is a dense area within the nucleus, which stains dark.

    b. The nucleolus is made up of RNA and proteins.

    c. The nucleolus is a membrane-bound area within the nucleus.

    d. The RNA and proteins making up the nucleolus are used in making ribosomes.

21. The ____________ is a fibrous, interconnected, protein framework that holds the cell together and anchors organelles in place.

22. The ___________ ___________ (two words) is known as the "packager" organelle of the cell.

23. Rough endoplasmic reticulum is studded with __________ and associated with ______ synthesis and production of the most common lipids, whereas smooth endoplasmic reticulum is associated with the production of ____________ in cells that make steroids or other wise specialize in the production of ___________.

    a. mitochondria, lipids, proteins, proteins

    b. mitochondria, proteins, lipids, lipids

    c. ribosomes, lipids, proteins, proteins

    d. ribosomes, proteins, lipids, lipids

24. Endocytosis includes:

    (i) taking particles into the cell by the process of _____________

    (i) enveloping small molecules by _____________

    (iii) capturing specific molecules in coated pits, a mechanism termed __________ -mediated endocytosis

25. Which of the following are functions of the cytoskeleton? List all that apply.

    a. Organization of the cell in space

    b. Support of organelles

    c. Cell motility

    d. Shape of the cell

    e. Scaffold for localizing the activity of macromolecules

26. Match:

List A	List B	
a. centrioles __	(i)	often anchors plasma membrane
b. microtubules __	(ii)	doublets, 9 + 2, many, short, in rows
c. basal body __	(iii)	flagella with a rotary motion
d. flagella __	(iv)	not found in (higher) plants
e. cilia __	(v)	made of tubulin subunits
f. actin filaments __	(vi)	triplets, 9+0, at "base"
g. intermediate filaments (fibers) __	(vii)	doublets, 9 + 2, few, long
h. bacterial flagellin (a protein) __	(viii)	helps cells resist mechanical stresses

For questions 27–31, answer true or false. If false, change the statement so as to make it true.

27. Vacuoles are common in animal cells, and contain water and dissolved substances. Vacuoles contribute to maintaining the shape and firmness (turgor) of cells.

28. Cytoplasmic streaming is due to myosin filaments moving over actin filaments and carrying chloroplasts or other organelles with them.

29. Exocytosis refers to the process of moving molecules into cells by fusing a vesicle with the plasma membrane.

30. Cell walls are found only in plants and prokaryotes.

31. Chloroplasts are found only in organisms in Kingdoms Plantae and Fungi.

32. Proteins and lipids are synthesized in the cell on endoplasmic reticulum (ER), and are moved within the ER or placed in membrane-bound vesicles derived from the ER. Where are these proteins and lipids next processed and what are some of the modifications that might be made to these molecules?

33. Lysosomes (sometimes called "spherosomes" in plants) are vesicles derived from the __________ and filled with _________ that can digest worn out organelles such as mitochondria as well as material taken in from the outside by endocytosis. Peroxisomes come from the _________ ER and break down molecules, generating potentially harmful ___________ ____________ (two words). This is degraded in turn into harmless substances such as oxygen or water.

34. Biological membranes are composed primarily of ____________ with embedded __________.

35. In the bilayer making up the cell (plasma) membrane, the interior of the bilayer is ___________ (hydrophobic / hydrophilic) and the exterior surfaces are ____________ (hydrophobic / hydrophilic). Experimental fusing of a mouse and a human cell, each with fluorescent-antibody labeled surface proteins, demonstrated that membrane proteins ________ (are / are not) free to move laterally within the membrane.

36. What are two differences that might be found between the interior and exterior sides of the plasma membrane of a cell?

37. Membrane proteins function in active or passive __________ , _____________ and cell ______________.

38. In a glass of table salt in water, the water is the ____________, the salt is the ___________, and the salt water is the ______________.

39. Diffusion is the movement of ____________ from a region of _________ concentration to a region of __________ concentration.

40. What is a concentration gradient?

41. Why do sugar molecules or other particles diffuse in a liquid?

42. What conditions are necessary for osmosis to occur?

43. What are some of the roles played by the proteins found in the plasma membrane?

44. The currently accepted model of the plasma membrane is the ________ ________ (two words) model.

45. The model referred to in question 44 was proposed in 1972 by

    a. Darnell and Lodish

    b. Singer and Nicolson

    c. Julianelle and Shia

    d. Davson and Danielli

46. A human red blood cell (rbc) has a concentration of solutes inside the cell equivalent to 0.9 percent NaCl. In each of the following cases, an rbc is placed in the solution described. Describe the movement of water across the cell membrane in each instance, and tell what will happen to the cell. Note that the cell membrane is impermeable to sodium and chloride ions, but freely permeable to water molecules.

    a. The rbc is placed in 3 percent NaCl solution (salt water).

    b. The rbc is placed in distilled water.

    c. The rbc is placed in 0.9 percent NaCl solution.

47. Refer to question 46. In each of the instances described, what would happen if a plant cell was substituted for the rbc? The plant cell in these cases also has an inside concentration of solutes equivalent to 0.9 percent NaCl. Compared to the inside of the cell, describe the solutions in a, b, and c as isotonic, hypertonic, or hypotonic.

48. Briefly describe the function of each of the following types of intercellular junctions, and indicate if the junction is found in animal, or plant cells.

    a. desmosomes, also called spot desmosomes

    b. tight junctions

    c. gap junctions

    d. plasmodesmata

49. The sodium-potassium pump establishes a(n) __________ gradient across the plasma membrane. For every _____ (one, two, or three) sodium ions transported ____ (out of, into) the cell, _____ (one, two, or three) potassium ions are moved ____ (out of, into) the cell. This pump requires energy derived from _____.

50. Complete the following table by filling in the blanks; the first line has been completed for diffusion as an example.

transport process	with or against the concentration gradient	requires energy from the cell	example(s)
diffusion	with	no	water, oxygen
______ transport	with	______	______
______ transport	______	yes	______

51. ________ ions are maintained in a higher concentration outside than inside the cell, and tend to diffuse back into the cell. The route the ions take back into the cell causes a monosaccharide molecule to be transported inside, along with the ion. This route is called a _________ channel.

52. The movement of hydrogen ions out of a cell involves what is called a ________ pump. As the protons diffuse back into the cell, their movement is coupled to the production of _____ from _____ and inorganic phosphate. This process is called ______________________.

53. Briefly define these terms:

    a. bulk flow

    b. diffusion

    c. tonicity

    d. osmosis

    e. water potential

54. What are some problems fish might face in fresh or salt water environments which are related to osmosis?

55. Which of the following ions or molecules can pass freely through the plasma membrane of cells? Why?

    Water, carbon dioxide, oxygen, hydrocarbons, glucose, amino acids, sodium ion, potassium ion, chloride ion.

56. What are some factors that influence the rate of diffusion?

Check Yourself

1. preexisting or earlier, fundamental or basic. **(Cells and cell theory)**

2. d. Robert Hooke **(Cells and cell theory)**

3. c. Schleiden and Schwann **(Cells and cell theory)**

4. b. As the size of a cell increases, the volume increases faster than the surface area. This means that, relative to increasing cell volume, there is a diminishing area for exchange of nutrients and wastes between the cell and its surroundings. **(Cells and cell theory)**

5. a. All cells share three characteristics: a cell or plasma membrane, cytoplasm, and genetic information stored in the form of DNA. Why are the other answers wrong? Plants, Monerans (bacteria) and fungi have cell walls, whereas animals cells lack a cell wall. In some viruses genetic information is stored in RNA, but in all cellular organisms DNA is used. Prokaryotes lack a true (membrane-bound) nucleus. **(Cells and cell theory)**

6. d. Most prokaryotic cells are between 1 and 10 micrometers in size, whereas eukaryotic cells generally range in size from 10 to 100 micrometers. An electron microscope must be used to see details of subcellular structures. **(Cells and cell theory)**

7. a. Answered in question as an example.

   b. Yes, prokaryotes and Eukaryotes both possess cell membranes.

   c. Prokaryotes and members of several eukaryotic kingdoms (plants, fungi, some protists) possess cell walls.

   d. Prokaryotes lack membrane-bound cell organelles such as mitochondria or vesicles.

   e. The single "chromosome" of prokaryotes is circular, and not complexed with proteins know as histones. It is not a chromosome in the sense of eukaryotic chromosomes. See "f."

   f. Circular chromosomes are characteristic of prokaryotes. Chloroplasts and mitochondria in eukaryotic cells have circular DNA, however, reflecting their probable origin as free-living prokaryotes.

   g. DNA is the genetic material for all organisms and viruses, with the exception of RNA viruses.

   h. Eukaryotic cells are subdivided by the endomembrane system (primarily made up of endoplasmic reticulum); prokaryotic cells are not subdivided in this manner.

   i. Ribosomes are smaller in prokaryotes than in eukaryotes. **(Cells and cell theory)**

8. The endosymbiotic theory suggests that mitochondria and chloroplast were originally free-living bacteria which entered the eukaryotic cell and over time evolved a mutually dependent relationship. Mitochondria

and chloroplasts are similar to bacteria in size, are double-membrane bound, and have their own circular DNA and small ribosomes like bacteria, and can produce some of their own proteins. The double membrane characteristic suggests one membrane is derived from a eukaryotic vesicle and the other from the bacterium. **(Mitochondria and chloroplasts)**

9. The outer membrane, inner membrane with cristae and the matrix. **(Mitochondria and chloroplasts)**

10. b. cellular respiration, energy, ATP, ADP (**Mitochondria and chloroplasts)**

11. True **(Eukaryotic cell organelles)**

12. True **(Eukaryotic cell organelles)**

13. Change "prokaryotic" to "eukaryotic." **(Mitochondria and chloroplasts)**

14. Change "mitochondria" to "chloroplasts." **(Mitochondria and chloroplasts)**

15. envelope, two. The nuclear envelope is a double membrane surrounding the nucleus. **(The nucleus)**

16. nuclear pores, proteins, into, out of. Nuclear pores are openings which selectively permit the passage of certain molecules. This channel role in membranes is always played by proteins embedded in the membrane (here termed a pore complex). Proteins need to be admitted from outside the nucleus since all proteins are synthesized outside the nucleus. RNA and RNA-containing molecules are made inside the nucleus and thus must be exported into the cytoplasm to function in protein synthesis. **(The nucleus)**

17. Grana, thylakoid, double **(Mitochondria and chloroplasts)**

18. cytoplasm, nucleoplasm **(The nucleus)**

19. nucleus, chromatin, chromosomes **(The nucleus)**

20. c. The nucleolus is not a membrane-bound structure. The nucleolus (or nucleoli—there may be several) is a region which stains darkly because of the ribosomal RNA (rRNA) and proteins collected there. **(The nucleus)**

21. cytoskeleton **(The cytoskeleton)**

22. Golgi apparatus **(Eukaryotic cell organelles)**

23. d. **(Eukaryotic cell organelles)**

24. (i) phagocytosis; (ii) pinocytosis; (iii) receptor **(Membrane structure and function)**

25. All of the items listed are functions of the cytoskeleton. **(The cytoskeleton)**

26. a. centrioles 4

    b. microtubules 5

    c. basal body 6

    d. flagella 7

    e. cilia 2

f. actin filaments 1

g. intermediate filaments (fibers) 8

h. flagellin (a protein) 3

Centrioles are not found in prokaryotes, fungi, or most plants. Centrioles are usually found near a microtubule organizing center, and can give rise to the spindle fibers (near the nucleus) or in the form of a basal body, form an anchor for flagella or cilia. Microtubules, actin filaments and intermediate filaments are all made up of protein subunits. Flagella and cilia are long and short cylinders composed of nine pairs of microtubules arranged around a central pair (9 + 2 array). Centrioles and basal bodies are cylinders of nine microtubule triplets, with none in the center (9 + 0 array). Bacterial flagella are made of long protein fibers, and rotate like a propeller. Actin filaments are often linked across the cell to proteins buried in the plasma membrane, and help hold the membrane and the cell together. Some centrioles, like mitochondria and chloroplasts, contain DNA and may have originally been free-living spiral-shaped bacteria which were taken into the cell. **(The cytoskeleton)**

27. False. Replace "animal cells" with "plant cells." Recall that animal and plant cells differ in several respects. Animals cells have centrioles, but lack cell walls, vacuoles, and chloroplasts or other plastids. **(Eukaryotic cell organelles)**

28. True **(The cytoskeleton)**

29. False. Replace "into cells" with "out of cells." **(Membrane structure and function)**

30. False. Cell walls are found in organisms in Kingdoms Plantae, Prokaryotae, Fungi, and Protista (Protoctista). **(Eukaryotic cell organelles)**

31. False. Fungi do not possess chloroplasts, and some protista (protoctists) have chloroplasts. "Chloroplasts are found only in organisms in Kingdoms Plantae and Protista (Protoctista)." **(Eukaryotic cell organelles)**

32. Proteins and lipids produced on the rough and smooth ER is shipped off to the Golgi bodies. With the Golgi, a protein might be made into a glycoprotein by the addition of a carbohydrate molecule; a lipid processed in this way would become a glycolipid. Other molecules might gain a phosphate group. The Golgi receive molecules made at other places in the cell, modify them, and package the finished compounds into membrane-surrounded bags, ready for shipment to another part of the cell or for secretion to the outside in secretory vesicles. **(Eukaryotic cell organelles)**

33. golgi, enzymes, smooth, hydrogen peroxide **(Eukaryotic cell organelles)**

34. phospholipids, proteins **(Membrane structure and function)**

35. hydrophobic, hydrophilic, are. The interior of the membrane bilayer is an oily mixture of the fatty acid ends of phospholipids. **(Membrane structure and function)**

36. On the exterior of the membrane, some cells have carbohydrate chains (glycolipid or glycoprotein) that act as recognition molecules. On the interior, there could be attachment points (membrane proteins) anchoring the plasma membrane to the rest of the cell's cytoskeleton. **(Membrane structure and function)**

37. transport, communication (as by hormones), recognition **(Membrane structure and function)**

38. solvent, solute, solution **(Diffusion, osmosis, and transport)**

39. particles, higher, lower **(Diffusion, osmosis, and transport)**

40. A concentration gradient exists in a volume of liquid when there is a higher concentration of molecules in one region that tapers off to a lower concentration in another region. For instance, imagine a sugar cube that has been at the bottom of a cup of coffee for a few minutes. The sugar concentration ranges from very concentrated at the sugar cube to less and less concentrated as one goes farther and farther away from the sugar cube. The concentration gradient will no longer exist when the sugar cube has totally dissolved and the concentration of sugar is the same throughout the coffee. This diffusion of sugar molecules is said to occur down the concentration gradient. **(Diffusion, osmosis, and transport)**

41. Molecules are always in motion at temperatures above absolute zero (minus 273 degrees Celsius). This thermal motion or agitation means molecules continually bang into each other. These collisions drive molecules from regions of higher concentration, where collisions are more frequent due to crowding, to regions of lower concentration where collisions are relatively infrequent. **(Diffusion, osmosis, and transport)**

42. For osmosis to occur, it is necessary to have (1) a semi-permeable membrane which is permeable to water but not to specific solute molecules (particles) such as sugars or proteins, and (2) a difference in concentration of these molecules that are dissolved in water on each side of the membrane. Under these conditions, diffusion will spontaneously occur. **(Diffusion, osmosis, and transport)**

43. Proteins can act as (1) receptors for specific molecules, (2) recognition (cell identity) molecules, (3) transport proteins or carriers for molecules or ions (e.g. sodium or hydrogen ions), (4) channels through which particles of smaller than the passageway can diffuse, (5) anchoring points tying the plasma membrane to the cytoskeleton, (6) adhesion molecules joining cell together, (7) enzymes promoting chemical reactions. **(Membrane structure and function)**

44. fluid mosaic **(Membrane structure and function)**

45. b. **(Membrane structure and function)**

46. a. The net movement of water will be out of the cell, from a region of higher water concentration (0.9% NaCl solution is 99.1 percent water) to a region of lower water concentration (3 percent NaCl solution is 97 percent water. Note that in this case the water diffuses (the NaCl can't get across the membrane) and that the diffusion of water occurs according to the same rule that applies to other particles—from a region of higher concentration to a region of lower concentration. As a result the cell shrinks and appears crumpled, a phenomenon termed crenation.

    b. The net movement of water will be into the cell, and the rbc will burst (cell lysis).

    c. There will be no net movement of water, since the concentration of water is the same on both sides of the cell membrane. That is, the concentrations of solutes and water are the same in and out of the cell, and the cell is said to be in osmotic equilibrium with its environment. It is important to understand that due to thermal motion, diffusion continues to occur. Water molecules continue to move across the membrane, but since the amount of movement is the same in both directions, on average, the state of equilibrium is maintained. This is called a dynamic equilibrium. **(Diffusion, osmosis, and transport)**

47. a. The plant cell would also lose water, like the rbc. Since a plant cell has a cell wall, however, the shape and size of the cell would remain about the same; the plasma membrane would pull away from the cell wall as water leaves the cell vacuole(s), a process called plasmolysis. The solution around the cell is hypertonic compared to the cell.

b. The net movement of water will be into the cell, swelling the vacuole, and pushing the plasma membrane against the cell wall. The solution around the cell is hypotonic compared to the cell. The plant cell is said to be turgid, which is the normal operatinig condition of a plant cell.

c. Water will move in and out of the plant cell since it is in an osmotic dynamic equilibrium with its environment. There will be no net movement of water. The surrounding solution is isotonic with the cell. **(Diffusion, osmosis, and transport)**

48. a. Demosomes bind animal cells together, in effect linking the cytoskeleton of one cell with that of an adjacent cell.

    b. Tight junctions, also found in animal cells, are like rivets, pulling cell membranes of adjoining cells together are points, and restricting the sideways movement of proteins within the plasma membrane.

    c. Gap junctions are communication pores between animal cells, like pipes or hollow rivets. These allow the movement of small ions and molecules between cells.

    d. Plasmodesmata are openings in the cell walls of adjacent plant cells. A strand of cytoplasm connects the two cells. **(Membrane structure and function)**

49. chemical (or electrochemical), three, out of, two, into, ATP **(Membrane structure and function)**

50. a. diffusion, only with, no, water, oxygen

    b. facilitated transport, only with, no, anions (in red blood cells)

    c. active transport, against or with, yes, amino acids and glucose

    Diffusion, osmosis and transport are three ways of moving particles across cell membranes. Two are based on simple physical phenomena, namely diffusion and osmosis. The other mechanisms are active transport and facilitated diffusion which require protein carriers and in the case of active transport, energy from the cell. Note that the only process that requires energy from the cell, active transport, is also the only mechanism for moving particles against (up) a concentration gradient, that is, from a region of lower concentration to a region of higher concentration. **(Diffusion, osmosis, and transport) (Membrane structure and function)**

51. sodium, coupled **(Membrane structure and function)**

52. proton, ATP, ADP, chemiosmosis **(Membrane structure and function)**

53. a. Bulk flow refers to the movement of a fluid in response to differences in pressure (water in a hose), gravity (a waterfall), or similar forces.

    b. Diffusion is the movement of particles from a region of higher concentration to a region of lower concentration.

    c. Tonicity is of way of indicating how the concentration of solute in one solution varies when compared to another solution. The terms of relative concentration are hypotonic, isotonic, and hypertonic.

    d. Osmosis is the movement of water through a selectively- or differentially-permeable membrane. The movement may be due to a difference in concentration of solute on each side of the membrane, or due to pressure exerted on the solution.

    e. Water potential is a concept that is sometimes used to explain or predict the direction of water movement. The "potential" referred to is the potential energy of the water. (Recall that energy is of two basic kinds: *potential energy,* or the energy associated with state or position; and *kinetic energy,* the energy of motion.) Water always moves, as would be expected, from a region of higher water potential to a region of lower water potential. Water behind a dam, e.g., is invested with higher water potential than water that has left the dam via a spill way, due to gravity. In a solution, the less water, the lower the water potential—a 5 percent sugar solution (95 percent water) has more potential than a 10 percent sugar solution (90 percent water). If a semipermeable membrane is placed between these two solutions, water will move from the region of higher water potential (95 percent water) to a region of lower water potential (90 percent water). This may also be viewed as the diffusion of water from an area of higher concentration to a region of lower concentration. **(Membrane structure and function)**

54. In freshwater, the tissues of the fish would be hypertonic compared to the surrounding medium, and the fish will tend to take up water. In a marine (saltwater) environment, the fishs tissues will be hypotonic compared to the water around it, and it will tend to lose water into the medium. **(Diffusion, osmosis, and transport)**

55. Small uncharged molecules such as carbon dioxide and oxygen freely diffuse across the plasma membrane, as do the uncharged and small polar molecules of water. Hydrocarbons and other hydrophobic molecules (such as organic solvents) can also freely penetrate the lipid bilayer. Ions, due to their charge can't get across the hydrophobic, uncharged region of the interior of the membrane. Polar molecules that are larger than water (molecular mass of 18) such as glucose (molecular mass of 180) or amino acids are likewise unable to cross the membrane on their own due to their size and polarity. **(Membrane structure and function)**

56. The rate of diffusion is affected by the size of the diffusing particles, temperature, distance, and particle concentration. Smaller particles diffuse faster than larger molecules (an atom of carbon-12 will diffuse faster than an atom of carbon-14), particles diffuse faster at higher temperature since there is faster movement of particles and thus higher-energy collisions, and particles diffuse much faster across a short distance than a long distance. Finally, at higher particle concentrations, there are more chances for particles to collide, and this contributes to particles spreading out faster. Also of importance is the area available through which diffusion can occur. A larger area of membrane will pass more particles in a given time than a smaller area, all other conditions being the same. **(Membrane structure and function)**

Grade Yourself

Circle the question numbers that you had incorrect. Then indicate the number of questions you missed. If you answered more than three questions incorrectly, you will then have to focus on that topic. (If a topic has less than three questions and you had at least one wrong, we suggest you study that topic also. Read your textbook, a review book, or ask your teacher for help.)

Subject: Cell Structure and Function

Topic	Question Numbers	Number Incorrect
Cells and cell theory	1, 2, 3, 4, 5, 6, 7	
Mitochondria and chloroplasts	8, 9, 10, 13, 14, 17	
Eukaryotic cell organelles	11, 12, 22, 23, 27, 30, 31, 32, 33	
The nucleus	15, 16, 18, 19, 20	
The cytoskeleton	21, 25, 26, 28, 29	
Membrane structure and function	24, 29, 30, 34, 35, 36, 37, 43, 44, 45, 48, 49, 50, 51, 52, 53, 55, 56	
Diffusion, osmosis, and transport	38, 39, 40, 41, 42, 46, 47, 50, 54	

Cell Energetics I

4

Brief Yourself

Chemical Reactions and Energy

Chemical reaction involve the making and breaking of bonds between atoms. Bonds are broken, new bonds formed, and energy released or absorbed. Endergonic reactions require and then store energy; exergonic reactions release energy.

Laws of Energy

Certain laws of energy (laws of thermodynamics) are known to govern everything that happens. The first law states, in simplified form, that energy cannot be created nor destroyed. This is the basis for the quantitative bookkeeping employed in studying the energetics of chemistry and life. Energy has to be accounted for if it appears or disappears in the course of a reaction. The energy "books" must balance in the end. The second law of energy may be cast in several forms. Basically, it states that as energy is changed from one form to another (e.g., the energy of the sun changed via photosynthesis into energy in carbohydrates), some energy is lost as heat. It is truly lost, since this heat energy is no longer available to do work.

Metabolic Pathways

Chemical reactions in organisms occur in a series of steps. One reason for this is so that energy can be liberated in small amounts that are compatible with life. (You can burn a sugar cube with a match or burn it stepwise, metabolically, by eating it—the former liberates energy much too rapidly for organisms to use safely.) These steps are linked together, the product of one serving as the reactant(s) of the next, in a series called a metabolic pathway. Metabolism is defined as the sum total of catabolic and anabolic processes. Catabolic processes release energy, while anabolic processes absorb (store) energy. One reaction (catabolic) may release energy used by another (anabolic) reaction. Such reactions are termed coupled reactions, referring to how they are tied together. For instance, a complex molecule may be broken down and the energy stored in a covalent bond between phosphate groups in the molecule ATP. This is one way of attaching phosphate groups—and their bond energies—to ADP.

Enzymes

Virtually all chemical reactions in the body require special chemicals called enzymes. Enzymes are catalysts: molecules which speed up chemical reactions but are unchanged at the conclusion of the reaction. That is, the reaction doesn't consume the enzyme, so it is available to speed up another reaction. They are generally complex proteins with fairly high molecular weights, and a very specific three-dimensional structure. An enzyme is specific in its action. The molecule an enzyme works on is called its substrate, and there is a particular substrate for every enzyme.

ATP Structure and Formation

ATP (adenosine triphosphate) is the universal energy currency used in cells. It is a way of banking energy for future use. ATP consists of an adenine base, a sugar, and three phosphate groups. The bonds between the second and third phosphate groups (and sometimes between the first and second) are made in storing energy and broken in giving up energy. Phosphorylation refers to the addition of a phosphate group to ADP (adenosine diphosphate). ATP can be made by two processes. In chemiosmotic phosphorylation a reservoir of protons is built up on one side of a membrane, and as they flow across the membrane, an ATP synthase complex uses the energy to attach a phosphate group to ADP. In the other form, substrate-level phosphorylation (as carried out in glycolysis), the phosphorylation is coupled to the breakdown of a complex organic molecule such as glucose.

Test Yourself

1. Define catalyst.

2. The substance on which an enzyme works is called its __________.

3. Which one of the following statements is not true of enzymes?

    a. Enzyme activity always increases as temperature increases.

    b. An enzyme functions best within a particular range of pH values.

    c. In almost all instances, enzymes are proteins.

    d. An enzyme is specific for a particular substrate.

4. The optimum temperature for most enzymes in the human body is about __ degrees Celsius.

5 An enzyme or class of enzymes is sometimes named after the substrate by combining the substrate name with the ending -_____. For example, an enzyme that breaks sucrose into fructose and glucose would be called _________.

6. According to the current view of enzyme action, an enzyme not only fits closely to its substrate, but the conformation of the enzyme is actually changed in forming the enzyme-substrate complex. This is called the ______ _______ (two words) model.

7. The energy of motion is called ________ energy. The energy of position or state is called ________ energy. Give an example of each.

8. Examine the following reaction:
lysine + alanine → lysine-alanine dipeptide + water

    a. Is this a hydrolysis or a condensation (dehydration synthesis) reaction?

    b. Where did the water come from?

    c. Would the reverse of this reaction be a hydrolysis or a condensation reaction?

For questions 9–13, indicate if the statement is true or false. If false, correct the statement to make it true.

9. The first law of energy (law of thermodynamics) states that energy can be neither created nor destroyed.

10. The second law of energy states that in converting energy from one form into another, some useful energy is always lost as heat.

11. An enzyme speeds up a chemical reaction in both directions, helping it reach chemical equilibrium faster.

12. In this reaction, the products are on the left and the reactants on the right:
hydrogen + oxygen + spark → water

13. In adding an inorganic phosphate group to ADP to make ATP, an ionic bond is created

between the second and third phosphate groups of the ATP molecule.

14. If an enzyme is heated excessively or subjected to strong acids or bases, it may be ____________, which means the molecule has undergone a conformational change and is no longer functional.

15. True or false: An enzyme molecule can only be used once in a chemical reaction.

16 What is the difference between a competitive and non-competitive inhibitor of enzymes?

17. In the process called glycolysis, a molecule of glucose is broken down to pyruvate in a series of enzyme-catalyzed reactions. This series of reactions is called a ________ pathway.

18. Examine the following reactions:
$A \rightarrow B \rightarrow C \rightarrow D \rightarrow E$

    As the concentration of the final product increases, it binds more and more often to the allosteric site on the enzymes catalyzing the reaction that changes A to B. This is an example of _______ __________ (two words).

19. Metabolism is composed of two parts, _________ and _________. In the former, substances are broken down in degradative biochemical pathways; in the latter, more complex and energy-rich compounds are synthesized.

20. True or false: All catalysts are proteins.

21. If a chemical reaction yields energy, it is called _________. If it requires (takes up) energy it is termed ________.

22. Name four factors which can affect the activity of an enzyme.

23. ATP can be made by two processes. One involves a proton pump and is called ____________________. In the other energy is transferred to ADP through the simultaneous degradation of an organic molecule, as in the cytoplasmic phase of the breakdown of glucose. This is called.______________

24. Most ATP is made by the process of ___________________.

25. If a chemical reaction that requires the input of energy derives that energy from another chemical reaction that proceeds at the same time, the reactions are said to be

    a. linked
    b. coupled
    c. organic
    d. endergonic

26. A system which energy and matter can enter or leave is termed __________, whereas a system that cannot exchange energy or material with its surroundings is termed _________.

27. The place on the enzyme molecule where it binds with the substrate is called the _______ site. The place where the inhibitor binds in non-competitive inhibition is the _________ site.

28. Label the following diagrams (where P stands for phosphate groups and A for the rest of the molecule) as representing ATP, AMP, or ADP.

    a. A - P - P
    b. A - P
    c. A - P - P - P

29. Co-enzymes are often associated with enzymes. Which of the following statements are true about co-enzymes?

    a. Co-enzymes include ions of calcium and magnesium.
    b. Co-enzymes include B-vitamins.
    c. Co-enzymes are proteins.
    d. Co-enzymes are non-protein molecules.

30. Enzyme cofactors include ions of elements such as

    a. magnesium
    b. sodium
    c. potassium

d. oxygen

31. Co-enzymes can have a role in oxidation-reduction (redox) reactions, meaning they function as ____________ acceptors or donors.

    a. proton

    b. hydroxyl group

    c. electron

    d. neutron

32. What is a calorie? What is a food calorie? How are joules related to calories?

33. In the following examples, indicate with an arrow the direction in which energy is moving. Example: "high energy low energy" becomes "high energy → low energy"

a.	hot object	cold object
b.	hot object	warm object
c.	slow moving particles	fast moving particles
d.	region of low organization (high disorder)	region of high organization (low disorder)

For questions 34–40. indicate if the statement is true of false. If false, correct to make it true.

34. Chemical reactions proceed in one direction only, from reactants to products.

35. A chemical reaction proceeds to the point where a specific ratio of products to reactants is reached. This point is called chemical equilibrium.

36. Some chemical reactions are irreversible.

37. The tendency for disorder to increase in a system or in the universe as a whole is called enthalpy.

38. Heat is due to the motion of atoms and molecules (particles).

39. A coenzyme is a nonprotein, non-organic molecule.

40. Contained within the structure of a molecule of NADH is a molecule of AMP.

41. How is energy stored in ATP?

42. a. What is the difference between NADH and NAD^+?

    b. NADH is a(n) ______________ .
    (possible answers: cofactor, coenzyme)

43. The role of NAD^+/NADH is to carry _________. (a. protons b. electrons c. neutrons)

44. FAD can accept two ___________ ions to become ____________.

45. The function of FAD is to carry __________. (a. protons b. electrons c. neutrons).

46. The ultimate source of energy for biological systems is the ________.

47. In making ATP by chemiosmotic phosphorylation, ____a____ are pumped across a membrane in the cell organelles ____b____ and ____b____. The high concentration of ____c____ on one side of the membranes causes an ____d____ gradient to be formed.

    Possible answers:

    For a: protons, electrons, neutrons

    For b: Golgi, chloroplasts, ribosomes, mitochondria

    For c: hydrogen ions, hydroxyl ions, oxygen

    For d: electronic gradient, electrochemical gradient

48. In making ATP by chemiosmotic phosphorylation, __________ _________ (two words) move from an area of high concentration to an area of low concentration, ___________ (down, up) an electrochemical ___________. This occurs across a membrane in association with a membrane bound complex, ATP ____________. The result is the synthesis of ATP.

Check Yourself

1. A catalyst is a substance that speeds up a chemical reaction without being permanently changed in the process. **(Enzymes)**

2. substrate **(Enzymes)**

3. a. Enzymes work best with a specific temperature range. Enzyme activity reaches a maximum at a point of optimum temperature, and falls off as the temperature rises above or falls below this point. If the temperature rises high enough, the enzyme will be denatured and no longer function as a catalyst. **(Enzymes)**

4. 37. This is average, or normal human body temperature. **(Enzymes)**

5. ase, sucrase **(Enzymes)**

6. induced fit. Enzyme molecules have an active (or binding) site where the enzyme and substrate join to form a short-lived enzyme-substrate complex. In this interaction, the shape (conformation) of the enzyme is actually changed. This is the induced fit model. **(Enzymes)**

7. kinetic, potential. An object falling or a chemical reaction giving off heat are examples of kinetic energy. The energy contained in a chemical bond between two atoms or in a car parked on a hill are examples of potential energy. **(Laws of energy)**

8. a. Condensation (dehydration synthesis)—a more complex molecule is being made.

   b. In forming the peptide bond between the two amino acids, one lost a hydrogen ion (H+) and the other an hydroxyl group (OH^-), which combined to form a molecule of water.

   c. The reversed reaction would be an example of hydrolysis, the breaking down of more complex molecules into simpler compounds or units, and involving the uptake of molecules of water. **(Chemical reactions and energy)**

9. True **(Laws of energy)**

10. True **(Laws of energy)**

11. True **(Enzymes)**

12. False. The reactants are on the lefthand side of the arrow, and the products on the right-hand side. **(Chemical reactions and energy)**

13. False. The bond is covalent, not ionic. **(Chemical reactions and energy)**

14. denatured **(Enzymes)**

15. False. An enzyme, like other catalysts, is not used up and may participate repeatedly in chemical reactions. **(Enzymes)**

16. A competitive inhibitor competes with the substrate for access to the binding or active site of the enzyme. By contrast, a non-competitive inhibitor affects the enzyme molecule by binding at the allosteric site. **(Enzymes)**

17. metabolic **(Metabolic pathways)**

18. feedback inhibition **(Metabolic pathways)**

19. catabolism, anabolism. Pathways may be synthetic, part of the building up process called anabolism (for instance making protein from amino acids). Or the pathways may be catabolic, breaking down complex materials into simpler components (turning a sugar cube into carbon dioxide and water). **(Metabolic pathways)**

20. False. RNA can act as an enzyme under special circumstances. Otherwise, enzymes are proteins. **(Enzymes)**

21. A chemical reaction which yields energy is called exergonic; if the energy is in the form of heat the term exothermic is often used. If a reaction requires an input of energy, it is termed endergonic. If the input is heat, the corresponding term is endothermic. **(Chemical reactions and energy)**

22. temperature, pH, the presence or absence of inhibitors, the concentration of enzyme, the concentration of substrate **(Conditions affecting enzymes)**

23. chemiosmotic phosphorylation, substrate level phosphorylation **(ATP—formation and structure)**

24. chemiosmotic phosphorylation **(ATP—formation and structure)**

25. b. coupled **(Chemical reactions and energy)**

26. open, closed **(Laws of energy)**

27. active (or binding), allosteric **(Conditions affecting enzymes)**

28. a. ADP

    b. AMP

    c. ATP **(ATP—formation and structure)**

29. b and d are true. **(Cofactors and enzymes)**

30. a. magnesium **(Cofactors and enzymes)**

31. c. electron **(Cofactors and enzymes)**

32. A calorie (sometimes called a chemical calorie) is the amount of energy needed to raise the temperature of one gram of water one degree Celsius. A large or food calorie (the dietary calories listed on the label of a box of food) is actually 1000 calories, or a kilocalorie. The Joule is another unit used to measure quantities of energy, and there are about 4 Joules in one (chemical) calorie. **(Chemical reactions and energy)**

33. a. hot object → cold object

    b. hot object → warm object

    c. slow moving particles ← fast moving particles

    d. region of low organization (high disorder) ← region of high organization (low disorder) **(Laws of energy)**

34. False. Chemical reaction can proceed in one direction only (irreversible reactions) but most proceed to form products and then reform reactants. **(Chemical reactions and energy)**

35. True **(Chemical reactions and energy)**

36. True **(Chemical reactions and energy)**

37. False. The tendency is called entropy. Enthalpy refers to the energy (heat) content of a substance. **(Laws of energy)**

38. True **(Laws of energy)**

39. False. A coenzyme is a nonprotein, organic molecule. **(Cofactors and enzymes)**

40. True **(Cofactors and enzymes)**

41. In the chemical bonds between phosphate groups. Energy is released by breaking these bonds, and stored by forming them. **(ATP - formation and structure)**

42. a. NADH is the reduced form. NADH has gained a hydrogen atom which is really a way of gaining an electron and the energy associated with that electron. b. cofactor **(Cofactors and enzymes)**

43. electrons. But remember what is seen is a gain or loss of hydrogen. **(Cofactors and enzymes)**

44. hydrogen, $FADH_2$. **(Cofactors and enzymes)**

45. electrons. **(Cofactors and enzymes)**

46. sun. As a broad generalization this statement is overwhelmingly correct. Deep in the ocean floor, however, there are geothermal vents which are the only known alternative to the sun as a source of energy for life on earth. **(Chemical reactions and energy)**

47. a. protons (i.e., hydrogen ions)

    b. chloroplast, mitochondria

    c. hydrogen ions (i.e. protons)

    d. electrochemical gradient **(ATP—formation and structure)**

48. hydrogen ions, down, gradient, synthase. **(ATP—formation and structure)**

Grade Yourself

Circle the question numbers that you had incorrect. Then indicate the number of questions you missed. If you answered more than three questions incorrectly, you will then have to focus on that topic. (If a topic has less than three questions and you had at least one wrong, we suggest you study that topic also. Read your textbook, a review book, or ask your teacher for help.)

Subject: Cell Energetics I

Topic	Question Numbers	Number Incorrect
Enzymes	1, 2, 3, 4, 5, 6, 11, 12, 13, 14, 20	
Chemical reactions and energy	8, 12, 13, 21, 25, 32, 34, 35, 36, 46	
Laws of energy	7, 9, 10, 26, 33, 37, 38	
Metabolic pathways	17, 18, 19	
Conditions affecting enzymes	22, 27	
ATP—formation and structure	23, 24, 28, 41, 47, 48	
Cofactors and enzymes	29, 30, 31, 39, 40, 42, 43, 44, 45	

Cell Energetics II

5

Brief Yourself

Cellular Respiration

Cellular respiration is the overall term given to the process of turning food such as glucose into ATP. One molecule of glucose can be metabolized into water and carbon dioxide with a net gain of about 36 ATP.

Mitochondria—Structure and Function

Mitochondria are the organelles of cellular respiration in the eukaryotic cell. Glycolysis happens in the cytoplasm outside the mitochondria, but the Krebs cycle occurs in the matrix (or inner compartment of the mitochodrion) and the electron transport system (ETS) is composed of sequential carriers grouped in the inner mitochondrial membrane.

An Overview of Glycolysis and Cellular Respiration

Glycolysis results in the breakdown of one molecule of glucose (which has six carbon atoms) into two molecules of pyruvate (which has three carbon atoms). The pyruvate is modified in the transition reaction into a 2-carbon compound, releasing carbon dioxide, and joins Coenzyme A (CoA) to enter the Krebs cycle, where additional carbon atoms are lost as carbon dioxide. ATP (sometimes in the form of GTP in the Krebs cycle), NADH and $FADH_2$ are the energy-rich products of glycolysis and the Krebs cycle. NADH and $FADH_2$ carry hydrogen atoms and their accompanying electrons into the ETS, where the majority of ATP is generated.

Substrate-Level and Oxidative Phosphorylation

ATP is produced in two ways. In substrate level phosphorylation, the phosphate group (and covalent bond energy) are derived from the substrate molecule, as happens in glycolysis when ATP is generated from intermediate compounds in the breakdown of glucose into pyruvate.

The other method of ATP production happens in the ETS. It is called oxidative phosphorylation and uses the membrane-based chemiosomotic process to make ATP .

Test Yourself

1. Glucose is a molecule with __________ carbon atoms. It can be completely metabolized (oxidized) to __________ and __________, with the accompanying release of energy.

2. Write the chemical equation for the complete oxidation of glucose.

3. The breakdown of glucose involves three major processes: ____________, the ____________ cycle and ___________ ___________ (three words). Energy will ultimately be obtained in the form of the universal currency of the cell, namely, ________.

4. Glycolysis involves the stepwise breakdown of one 6-carbon glucose molecule to two 3-carbon molecules of __________.

5. The step linking glycolysis to the Krebs cycle is the ____________ step.

6. The Krebs cycle is the source for the ________ __________ we exhale.

7. The net production of ATP for glycolysis is _______, for the Krebs cycle is _____, and for the ETS is ______, for a total net production of _____ ATP.

   a. 2, 4, 16, 22

   b. 4, 4, 30, 36

   c. 2, 2, 32, 36

   d. 4, 2, 32, 38

8. Where do these processes happen in the cell? Glycolysis occurs in the _________, the Krebs cycle in the _____________, and the ETS in the ______________.

9. Fermentation occurs in the absence of ______________.

10. In yeast the products of fermentation are ________ and __________. In your muscles, the products of fermentation are __________ and _____________.

11. In the first step of glycolysis, ATP is used to add a ___________ ___________ (two words) to glucose. This step is called __________________.

12. The next step is a rearrangement of glucose-6-phosphate to __________-6-phosphate (hint: this sugar is an isomer of glucose), followed by the addition of a second ________ ________ (two words).

13. The _____ (number) -carbon sugar is split into two ____ (number) -carbon molecules in the next step of glycolysis.

14. In the next steps, NAD+ is made into _________, and ATP is also made, the latter requiring the loss of ___________ groups. These steps are repeated for each ___ (number) -carbon molecule.

15. Finally, the removal of a second __________ group from phosphoenolpyruvate results in the formation of _______________.

16. In the transition reaction, the pyruvate dehydrogenase complex of enzymes removes _____________ (one or two words) from pyruvate, making the _____ -carbon pyruvate compound into _____ -carbon __________ molecule which combines with Coenzyme A to make ____________.

    a. oxygen, 2, 3, acetyl, acetyl-CoA

    b. oxygen, 3, 2, acetyl, acetyl-CoA

    c. carbon dioxide, 3, 2, acetyl, acetyl-CoA

    d. carbon dioxide, 2, 3, acetyl, acetyl-CoA

17. What is another name for the Krebs cycle?

18. What are the products of the Krebs cycle?

19. The Krebs cycle begins with the 2-carbon acetyl-CoA molecule combining with a ___

(number) -carbon oxaloacetate molecule to yield a ___ (number) -carbon citrate molecule.

20. In the next steps, two molecules of _________ _________ (two words) are removed from isocitrate to give ____ (number) -carbon molecule of succinyl CoA.

21. When Succinyl CoA is transformed, an energy-rich compound is made called ___________.

22. Which of these processes are aerobic? Which are anaerobic?

    (i) glycolysis

    (ii) Krebs cycle

    (iii) Electron transport system

23. What is passed from glycolysis and the Krebs cycle to the electron transport system (ETS)?

24. What compounds are used in the ETS to transfer energy from electrons to ATP?

25. What is the final hydrogen acceptor in the ETS. What product is formed as a result?

26. Is ATP stored in large quantities in the cell, or are smaller quantities used and continually replenished?

27. How are lipids used in cells to supply energy?

28. How are amino acids and proteins used as energy sources?

29. What is the caloric value of:

    (i) one gram of carbohydrate

    (ii) one gram of lipid

    (iii) one gram of protein

30. Define autotroph.

31. What is a heterotrophic organism?

32. How do electrons enter the ETS?

33. How could you tell by microscopic examination if a cell uses a lot of energy (e.g., muscle cells, or cells that synthesize a lot of materials, as in the pancreas)?

34. Briefly compare glycolysis and cellular respiration to photosynthesis.

35. Where in the mitochondrion do each of these occur?

    (i) glycolysis

    (ii) Krebs cycle

    (iii) Electron transport system

36. Sketch a mitochondrion, showing the (a) outer membrane (b) matrix and (c) cristae.

37. True or false: A mitochondrion is much smaller than a eukaryotic cell, about the size of a bacterial cell.

38. How is ATP generated within the mitochondrion?

39. In what part of the mitochondrion is the pH lowest? What is the significance of this fact?

40. How does ATP leave the matrix?

41. FAD and $FADH_2$ are functionally analogous to _________ and _________.

42. In glycolysis, ____ (number) molecule of glucose is turned into _______ (number) molecules of pyruvate.

43. For each molecule of pyruvate that leaves glycolysis and enters the Krebs cycle, the following numbers (fill in the blanks) of molecules are produced: ___ molecules of $FADH_2$, ___ molecules of ATP (GTP), ___ molecules of carbon dioxide, and ___ molecules of NADH.

44. How many turns of the Krebs cycle are required to process one molecule of glucose?

45. How does cyanide kill?

Check Yourself

1. 6, CO_2, H_2O **(Overview of glycolysis and cellular respiration)**

2. $C_6H_{12}O_6$ + 6 CO_2 ———————- 6 CO_2 + 6 H_2O **(Overview of glycolysis and cellular respiration)**

3. glycolysis, Krebs, electron transport system, ATP **(Overview of glycolysis and cellular respiration)**

4. pyruvate **(Overview of glycolysis and cellular respiration)**

5. transition. Pyruvate enters the Krebs cycle after being converted into a 2-carbon molecule and joined to a molecule of CoA. **(Overview of glycolysis and cellular respiration)**

6. carbon dioxide **(Overview of glycolysis and cellular respiration)**

7. c. Glycolysis requires ATP, so the net gain of ATP is only two molecules per molecule of glucose processed. The overall energy gain per molecule of glucose, from two turns of the Krebs cycle, is 2 ATP. The ATP production in the ETS is 32 ATP, using NADH (from glycolysis and the Krebs cycle) and $FADH_2$ (from the Krebs cycle). **(Overview of glycolysis and cellular respiration)**

8. cytoplasm, matrix, inner membrane **(Overview of glycolysis and cellular respiration)**

9. oxygen. Glycolysis is an anaerobic process in that free oxygen is not required. The Krebs cycle and ETS are aerobic, however, and if free oxygen is not available, fermentation occurs instead. The drawback to fermentation is that the energy yield from fermentation is much less than the yield through the Krebs cycle and ETS. **(Overview of glycolysis and cellular respiration)**

10. carbon dioxide, alcohol, carbon dioxide, lactate (lactic acid) **(Glycolysis)**

11. phosphate group, phosphorylation **(Glycolysis)**

12. fructose, phosphate group. The second phosphate makes the product fructose-6-diphosphate. **(Glycolysis)**

13. 6, 3 **(Glycolysis)**

14. NADH, phosphate groups, 3. That is, the phosphate groups originally attached to the sugars are transferred when making ATP. **(Glycolysis)**

15. phosphate, pyruvate. Overall, in glycolysis, glucose is manipulated in a series of steps to first form a glucose-phosphate molecule, glucose-6-phosphate. Then it is turned into fructose-6-phosphate, another phosphate group is added and further processing occurs. The glucose molecule is split into two phosphoglyceric acid (PGA) molecules and finally processed into pyruvate (sometimes called pyruvic acid). This is an anaerobic process in which free oxygen is not required. **(Glycolysis)**

16. c. carbon dioxide, 3, 2, acetyl, acetyl-CoA **(Glycolysis)**

17. TCA cycle (tricarboxylic acid cycle), or citric acid cycle **(Krebs cycle)**

18. The products of the Krebs cycle are carbon dioxide, GTP (ATP), $FADH_2$, and NADH **(Krebs cycle)**

19. 4, 6 **(Krebs cycle)**

20. carbon dioxide, 4 **(Krebs cycle)**

21. ATP. (GTP is actually a more correct answer.) **(Krebs cycle)**

22. (i) anaerobic (ii) aerobic(iii) aerobic **(Krebs cycle)**

23. NADH, $FADH_2$. NADH come from glycolysis and the Krebs cycle and $FADH_2$ from the Krebs cycle. **(Krebs cycle)**

24. cytochromes. In the electron transport system, a series of molecules called cytochromes (similar to the molecules used in electron transport in photosynthesis) move electrons along a series of oxidation-reduction reactions. At each step, energy is removed from the electrons to make ATP. **(Electron transport system)**

25. oxygen, water. The final electron acceptor is oxygen which ultimately forms water. **(Electron transport system)**

26. Smaller quantities of ATP exist in the cell and rapidly turned over. Many molecules of ATP are synthesized each day in a cell. **(Electron transport system)**

27. Lipids are first broken down into their components, glycerol and fatty acids. The glycerol can enter glycolysis in the form of PGAL, and processed like glucose from that point on. Fatty acids are metabolized and enter the cell respiration system at the transition reaction, as acetyl in acetyl CoA. **(Electron transport system)**

28. Proteins are broken down into amino acids. The mono acids are deaminated (nitrogen removed) forming nitrogenous waste. Depending on the amino acid, the carbon backbone is fed into the glycolysis-cell respiration process as pyruvate, acetyl, or into the Krebs cycle in the form of another intermediate compound. **(Electron transport system)**

29. One gram of carbohydrate supplies 4 kilocalories of energy. One gram of protein supplies 4 kilocalories. One gram of fat, however, supplies 9 kilocalories. Note that the kilocalorie (1,000 calories) is equivalent to one food or dietary calorie. **(Overview of glycolysis and cellular respiration)**

30. An autotroph is an organism, such as a green plant, that needs only simple inorganic compounds and a source of energy (such as sunlight) in order to live. **(Overview of glycolysis and cellular respiration)**

31. Heterotrophic organisms require a source of more complex organic compounds for energy, which are not required by autotrophs. **(Overview of glycolysis and cellular respiration)**

32. The ETS obtains electrons from NADH and $FADH_2$. The electrons are attached to the hydrogen atoms picked up by NAD+ and FAD molecules in glycolysis and the Krebs cycle. Since gaining electrons is the process of reduction, the hydrogen and accompanying electrons are sources of reducing power for the ETS. Or stated more technically, as the energetic electrons pass through the ETS, their free energy levels diminish and by the end, they are greatly reduced. **(Overview of glycolysis and cellular respiration)**

33. A cell that uses energy in large amounts will have many mitochondria. **(Mitochondria—structure and function)**

34. Glycolysis and cellular respiration break glucose down into simpler compounds (water and carbon dioxide) and release energy in the process. Through photosynthesis, green plants use carbon dioxide, water and solar energy to make more complex molecules (carbohydrates). In terms of the end result, if not

the individual biochemical steps, glycolysis and cellar respiration are the reverse of photosynthesis. **(Overview of glycolysis and cellular respiration)**

35. (i) Glycolysis occurs in the cytosol, not in the mitochondrion.

    (ii) The Krebs cycle is located within the matrix.

    (iii) The ETS is located on the inner mitochondrial membrane. **(Mitochondria—structure and function)**

36.

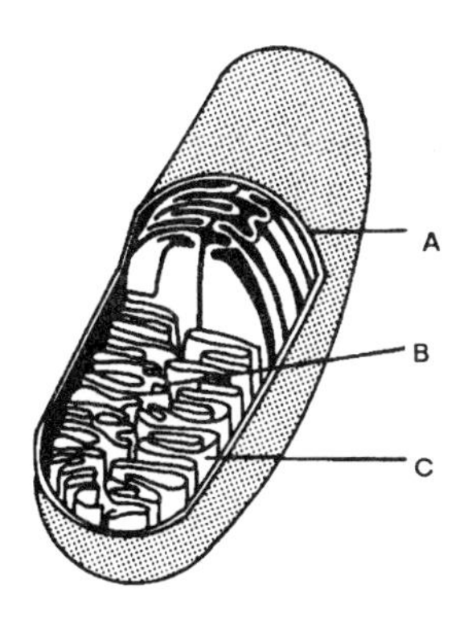

(Mitochondria—structure and function)

37. True. This fact is consistent with the endosymbiotic theory of the origin of mitochondria. **(Mitochondria—structure and function)**

38. ATP is generated in the mitochondrion indirectly in the Krebs cycle (substrate-level phosphorylation), and directly in the ETS (chemiosmotic synthesis of ATP). **(Substrate-level and oxidative phosphorylation)**

39. The pH in the mitochondrion will be lowest where the concentration of protons is highest (recall the definition of pH). The more acid compartment is the outer compartment between the inner and outer membranes. The high concentration of protons creates an electrochemical gradient across the membrane; protons spontaneously diffuse into the matrix, driving the formation of ATP by the ATP synthase complex. **(Substrate-level and oxidative phosphorylation)**

40. ATP leaves the mitochondrion via a protein channel. **(Substrate-level and oxidative phosphorylation)**

41. NAD+, NADH **(Krebs cycle)**

42. 1, 2 **(Glycolysis)**

43. 1, 1, 2, 3 **(Krebs cycle)**

44. 2 **(Krebs cycle)**

45. Cyanide reacts with one of the cytochrome molecules in the electron transport chain, and prevents the electrons from being transferred to oxygen (the electron acceptor). **(Krebs cycle)**

Grade Yourself

Circle the question numbers that you had incorrect. Then indicate the number of questions you missed. If you answered more than three questions incorrectly, you will then have to focus on that topic. (If a topic has less than three questions and you had at least one wrong, we suggest you study that topic also. Read your textbook, a review book, or ask your teacher for help.)

Subject: Cell Energetics II

Topic	Question Numbers	Number Incorrect
Overview of glycolysis and cellular respiration	1, 2, 3, 4, 5, 6, 7, 8, 9, 29, 30, 31, 32, 34	
Glycolysis	10, 11, 12, 13, 14, 15, 16, 42	
Krebs cycle	17, 18, 19, 20, 21, 22, 23, 41, 43, 44, 45	
Electron transport system (ETS)	24, 25, 26, 27, 28	
Mitochondria—structure and function	33, 35, 36, 37	
Substrate-level and oxidative phosphorylation	38, 39, 40	

Photosynthesis

6

Brief Yourself

Light

Visible light is part of the much larger electromagnetic spectrum that includes radio waves, microwaves, infrared, ultraviolet, and X-rays. Visible light appears white to our eyes, but can be decomposed using a prism into its constituent wavelengths seen as the colors red, orange, yellow, green, blue, indigo, and violet.

Overview of Photosynthesis

Photosynthesis is the process of absorbing light energy from the sun through the use of special pigment molecules of chlorophyll (which comes in *a, b*, and other forms) or carotenoids. The energy is used to boost electrons derived from water to a higher energy state in Photosystem II (PS II). In the cyclic electron pathway, the electrons are then passed along an electron transport chain where the energy is slowly bled off, step-by-step, to make ATP. The electrons are re-energized by solar energy in Photosystem I (PS I) and then added to $NADP^+$ (along with protons) to make NADPH. The ATP and NADPH are used in the Calvin cycle, where carbon dioxide is fixed and invested with energy to make a 3-carbon molecule that can be used to make carbohydrates, proteins, and lipids.

Chloroplast—Structure and Function

A plant leaf contains photosysnthetic cells, each with many chloroplasts. The chloroplast is the organelle of photosynthesis in eukaryotes. It is a double-membrane structure, containing a fluid called stroma and stacks (grana) of hollow sacs called thylakoids. The light-dependent reactions (PS I and PS II) occur in the thylakoid membrane. The light-independent reactions (Calvin cycle) occur in the stroma.

C3, C4, and CAM Plants

Not all plants use the same system for grabbing onto carbon dioxide. The carboxylating enzyme associated with the ordinary C_3 process, Rubisco (1,5-ribulose bisphosphate carboxylase), is inefficient at elevated temperatures. C_4 and CAM plants avoid this limitation by using an enzyme called PEPCO. (C_3 and C_4 refer to the number of carbons formed in the first stable products; CAM stands for the crassulacean-acid metabolism of succulent plants.)

Test Yourself

1. Light is a form of ___________ radiation, along with X-rays, microwaves, and radio waves. All of these are made up of particles called __________.

2. Electromagnetic radiation may be viewed as either a ______________ (like a baseball) or a ______________ (as in the ocean).

3. Different forms of electromagnetic radiation have different amounts of energy. The _________ (longer or shorter) wavelength radiation like radio waves has ________ (more or less) energy than the shorter wavelength radiation such as X-rays.

4. Arrange these types of electromagnetic radiation in order of increasing energy, that is from the weaker (less energetic photons) to the stronger (more energetic photons).

   X-rays, red light, blue light, ultra-violet, infra-red, radio waves

5. List the colors of the spectrum, starting from the end nearest infra-red light.

6. Objects can transmit, absorb or reflect light. A leaf which reflects red light, absorbs blue light, and transmits other wavelengths will appear to be what color to the human eye?

7. Chloroplast, mitochondria, and nuclei are all unique in that they have a ________ (single/double/triple) outer membrane.

8. Inside the chloroplast, there are stacks of hollow, membranous sacs that look like piled pancakes. The sacs are the ___________, and the stacks are ___________ . The singular form of the latter word is __________.

   a. grana, thylakoids, thylakoid

   b. thylakoids, grana, granum

   c. thylakoids, granum, grana

   d. stroma, grana, granum

9. The compartment in which the grana are contained is filled with a dense, enzyme-rich solution, the ___________.

Questions 10–18. Matching.

Possible answers (used only once, but not all used): Photosystem I, Photosystem II, fixation of carbon dioxide, production of oxygen, NADPH and ATP, ATP synthase, stroma, inside the thylakoid, NADPH, Calvin cycle, light-dependent reactions.

10. These reactions happen on the thylakoid membrane

11. Cycle that occurs in the stroma

12. These carry energy and reducing power from the light-dependent reactions to the Calvin cycle

13. At the beginning of the Calvin cycle

14. $NADP^+$ plus a proton

15. Makes ATP

16. Protons concentrated here, forming an electrochemical gradient across the membrane

17. P_{700} - which photosystem?

18. P_{680} - which photosystem?

19. Pick the answer that is WRONG. The pigments involved in photosynthesis include:

    a. chlorophyll *a*

    b. chlorophyll *b*

    c. chlorophyll *c*

    d. carotenoids

    e. cytochrome C

20. For the correct answers in question 19,

    a. List the pigment(s) that is (are) green:

    b. List the pigment(s) that is (are) red, orange, or yellow:

21. In the __________-dependent reactions, _________ energy is captured, and ________ is split releasing oxygen and supplying electrons to replace those lost from chlorophyll *a* molecules.

22. The oldest photosystem is Photosystem I, which has a reactive, central molecule chlorophyll ______ (*a* or *b*), referred to as _________ (P_{680} or P_{700}).

23. Photosystem I involves the ______ (cyclic or noncyclic) electron pathway and produces:

    a. ATP

    b. Protons, electrons, and oxygen from water; ATP; NADPH

24. The youngest photosystem is Photosystem II, which has as a reactive, central molecule chlorophyll ______ (*a* or *b*), referred to as _________ (P_{680} or P_{700}).

25. Photosystem II involves the ______ (cyclic or noncyclic) electron pathway and produces:

    a. ATP

    b. Protons, electrons, and oxygen from water; ATP; NADPH

26. In each photosystem, chlorophyll and carotenoid pigment molecules are tightly packed, and form ___________ to gather light energy and channel it into the reaction center made of chlorophyll ___. (Pick two answers from those below.)

    a. *a*

    b. *b*

    c. antennas

    d. NADPH

    e. photons

27. The ___________ spectrum consists of the wavelengths of light necessary for photosynthesis.

    a. action

    b. emission

    c. absorption

    d. adsorption

28. Green plants use blue and red light in photosynthesis, but as anyone can see, sunlight is white. Where do the colors come from?

29. The __________ (cyclic, noncyclic) electron pathway generates only ___ and recycles electrons as part of Photosystem ____ (I or II).

30. There are four mistakes in this statement—correct them. (Assume the statement does refer to the cyclic pathway.) The cyclic electron pathway produces a number of products, including the energy rich compound NADPH and the electron carrier ATP. Electrons for this electron pathway are derived from splitting water, which also produces nitrogen and protons.

31. Place the correct labels for the items assigned letters A-G.

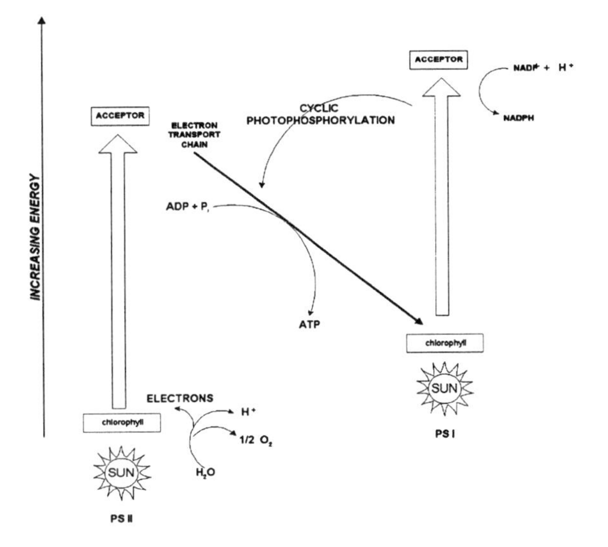

32. In turning $NADP^+$ into NADPH, how many electrons are needed, one, two or three?

33. In splitting water into electrons, protons, and oxygen, how many protons and how many electrons are produced?

34. Where do the protons produced by the splitting of water go? What are they used for?

35. Where do the light-dependent reactions take place?

36. Why is the phrase "light-independent reactions" more accurate than "dark reactions"?

37. Who received a Nobel prize in 1961 for work on the light-independent reactions? What radioisotope did he and his colleagues use in their research?

38. What does RuBP stand for? What is Rubisco (sometimes, RubisCO)?

39. Where does carbon dioxide necessary for photosynthesis enter the plant?

40. One way of understanding the Calvin cycle is to break it down into three stages. What are the three stages?

41. What is the molecule that accepts the carbon dioxide molecule in the Calvin cycle? How many carbon atoms are in this molecule?

42. How are Photosystems I and II connected?

43. What is the first product in the carbon dioxide reduction part of the Calvin cycle? How many carbon atoms does it contain?

    a. PGAL

    b. PGA

    c. PGAP

    d. RuBP

44. What is the next product in the carbon dioxide reduction part of the Calvin cycle? How many carbon atoms does it contain?

    a. PGAL

    b. PGA

    c. PGAP

    d. RuBP

45. What product is regenerated into RuBP?

    a. PGAL

    b. PGA

    c. PGAP

46. What is the product of the Calvin cycle that can be used to produce carbohydrates and other materials needed by the plant?

47. What is the source of energy in the Calvin cycle? Where does this come from?

48. What is the source of reducing power in the Calvin cycle? Where does it come from?

49. Distinguish between C_3, C_4, and CAM plants in terms of their leaf anatomy.

50. What is photorespiration and how does it affect photosynthesis? Under what environmental conditions is photorespiration a problem?

51. True or false: All energy from the sun that reaches the earth's surface is used in photosynthesis.

52. What are the two basic strategies used by plants to avoid the problem of photorespiration? Plants using PEPCO are not as productive as Rubisco-using plants at cooler temperatures. Why, then, do some plants use PEPCO?

53. What plants use PEPCO to avoid photorespiration?

Check Yourself

1. electromagnetic, photons. Light has characteristics of both particles (like baseballs or bullets) and waves (like ocean waves). Treated as a particle, light is composed of small packets of energy called photons; the more energetic forms of electromagnetic radiation are made up of higher energy photons. Light may also be discussed in terms of wavelengths, and numbers assigned to different wavelengths. Shorter wavelengths of light have more energy, and *vice versa.* **(Light)**

2. particle (photon), wave **(Light)**

3. longer, less **(Light)**

4. radio waves, infra-red (IR), red, blue, ultra-violet (UV), X-rays **(Light)**

5. (infra-red) red, orange, yellow, green, blue, indigo, violet (ultra-violet). In this order, the first letter of each of the colors can be recalled as a name, "ROY G BIV." **(Light)**

6. The leaf would be seen as red, as the red wavelengths would be reflected from the leaf to the viewer's eye. The blue light is absorbed and not seen, and the transmitted light isn't seen (on that object). Green plants make use of only part of the visible light spectrum—mainly red and blue. The plants appear green because they reflect (to our eyes) the green wavelengths of light. We see reflected light, but do not see light that is absorbed or transmitted through objects. **(Light)**

7. double **(Chloroplast—structure and function)**

8. b. thylakoids, grana, granum **(Chloroplast—structure and function)**

9. stroma **(Chloroplast—structure and function)**

10. light-dependent reactions (also could be answered as photosystem I and photosystem II) **(Overview of photosynthesis)**

11. Calvin **(Overview of photosynthesis)**

12. NADPH and ATP are produced in the light-dependent reactions and fed into the Calvin cycle. **(Overview of photosynthesis)**

13. fixation of carbon dioxide **(Overview of photosynthesis)**

14. NADPH (also carries two electrons) **(Overview of photosynthesis)**

15. ATP synthase (complex) makes ATP via chemiosmosis, using the protons stored in the thylakoids. **(Overview of photosynthesis)**

16. Inside of the thylakoid. **(Overview of photosynthesis)**

17. Photosystem I **(Overview of photosynthesis)**

18. Photosystem II **(Overview of photosynthesis)**

19. Cytochrome C **(Light-dependent reactions)**

20. a. Green pigments are chlorophylls *a* and *b*.

    b. Carotenoids **(Overview of photosynthesis)**

21. light, solar (sunlight), water **(Light-dependent reactions)**

22. chlorophyll *a*, P_{700} **(Light-dependent reactions)**

23. cyclic, a **(Light-dependent reactions)**

24. a, P_{680} **(Light-dependent reactions)**

25. noncyclic, b **(Light-dependent reactions)**

26. antennas (antennae), *a* **(Light-dependent reactions)**

27. a is the best answer, although c might be another possible answer. **(Overview of photosynthesis)**

28. Sunlight is a mixture of different wavelengths of light. Together they appear white, but as Newton showed in the 17th century, passing sunlight through a prism breaks it down into its component wavelengths; passing these back out through a prism gives white light once again. **(Light)**

29. cyclic, ATP, Photosystem I **(Light-dependent reactions)**

30. The correct statement is: The noncyclic electron pathway produces a number of products, including the energy rich compound ATP and the electron carrier NADPH. Electrons for this electron pathway are derived form splitting water, which also produces oxygen and protons. This set of reactions is best summarized diagrammatically, in the *z-scheme* (as shown in question 31). Electrons are passed from PS II to PS I along an electron transport chain where ATP is made, and finally used to make NADPH (not NADH). This is thc noncyclic electron pathway. In the cyclic pathway, electrons are moved around PS I and recycled, and used to make ATP. **(Light-dependent reactions)**

31. (A) cyclic photophosphorylation (B) ATP (C) PSI (D) PSII (E) H^+ (F) NADH (G) H^+ **(Light-dependent reactions)**

32. two **(Light-dependent reactions)**

33. Two protons and two electrons are produced per molecule of water split. Not all photosynthetic organisms use water as a source of electrons. Some bacteria use hydrogen (from hydrogen or hydrogen sulfide) in place of water. Such bacteria do not produce oxygen as a by-product of photosynthesis. **(Light-dependent reactions)**

34. The protons accumulate inside the thylakoids, and are used in the chemiosmotic synthesis of ATP. **(Light-dependent reactions)**

35. In the thylakoid membrane **(Light-dependent reactions)**

36. While the light-independent reactions can and do occur in the dark, the reactions also occur in the presence of light. "Light-independent reactions" is a more descriptive phrase, indicating that light is not necessary for the reactions to take place. **(Light-independent reactions)**

37. Melvin Calvin and his coworkers used carbon-14 radioisotope. **(Light-independent reactions)**

38. "RuBP" is the abbreviation for ribulose bisphosphate, a 5-carbon "carrier" for carbon dioxide. "Rubisco" is the enzyme RuBP carboxylase (note the "carboxyl-" in the name of the enzyme, indicating it catalyzes the addition of a molecule of carbon dioxide to RuBP). **(Light-independent reactions)**

39. Carbon dioxide enters the plant through stomata (singular: stoma), openings found on leaves and stems. These openings are controlled by guard cells which can open or close. **(Light-independent reactions)**

40. The three stages are: (i) carbon dioxide fixation in which carbon dioxide is incorporated into the RuBP molecule to make a 6-carbon molecule; (ii) the temporary 6-carbon molecule is split into two 3-carbon molecules of PGA, and further modified into PGAP with energy from ATP and reducing power from NADPH; (iii) RuBP is regenerated for another round, by making 3 RuBP molecules (each with 5 carbon atoms) from 5 molecules of PGAL (each with 3 carbon atoms). These reactions occur in the stroma. **(Light-independent reactions)**

41. Carbon dioxide is accepted into the Calvin cycle by RuBP, a 5-carbon molecule. **(Light-independent reactions)**

42. PS. I is connected to PS II by an electron transport system. Two electrons are transported down this "electronic staircase" for every one molecule of ATP that is generated. Energy from high-energy electrons—boosted to this level by the energy from the sun—is bled off in a stepwise fashion. **(Light-dependent reactions)**

43. b. Although an unstable 6-carbon molecule is formed first, technically speaking, the first stable product is the 3-carbon molecule, PGA. **(Light-independent reactions)**

44. c. PGAP is the next product, and is also a 3-carbon molecule. **(Light-independent reactions)**

45. a. PGAL is regenerated into RuBP. **(Light-independent reactions)**

46. PGAL is the primary product of the Calvin cycle, a raw material from which the plant can synthesize other carbohydrates, proteins and lipids. **(Light-independent reactions)**

47. The Calvin cycle derives energy from ATP produced in the light-dependent reactions. **(Light-independent reactions)**

48. The source of reducing power is NADPH, produced in the light-dependent reactions. **(Light-independent reactions)**

49. The C_3 plants have bundle sheath cells without chloroplasts; C_4 plants have bundle sheath cells with chloroplasts; and CAM plants have large vacuoles in their mesophyll cells. **(C3, C4 and CAM plants)**

50. Photorespiration occurs when carbon dioxide levels in the leaf drop, and are lower relative to the amount of oxygen present. The problem is that Rubisco can act as a carboxylase or oxygenase. Oxygen competes with carbon dioxide for the binding site of the Rubisco enzyme molecule. When oxygen reacts with RuBP, it forms PGA and phosphoglycolate, and finally, carbon dioxide is reformed. This circular process wastes energy, since there is no net gain of anything. Photorespiration is most likely to occur in C_3 plants under hot, dry conditions. The stomata are closed to conserve water, but this means there is no additional carbon dioxide available. **(C3, C4 and CAM plants)**

51. False. Only a small proportion (less than 1 percent) of solar energy reaching the earth's surface is used in photosynthesis. **(Overview of photosynthesis)**

52. C_4 and CAM plants both use phosphoenolpyruvate carboxylase (PEPCO, or sometimes, PEPCase) in place of Rubisco. PEPCO does not have Rubisco's affinity for oxygen, and so avoids the basic reaction of photorespiration. In C_4 plants, the supply of carbon dioxide is fixed as oxaloacetate in mesophyll cells and passed on to bundle sheath cells, where the Calvin cycle is carried out. This is often referred to as "partitioning in space" meaning that the process of taking up carbon dioxide is physically (spatially) separated from the Calvin cycle. CAM plants, on the other hand, "partition in time", that is, they take up carbon dioxide into a 4-carbon molecule during the night (when stomata can be wide-open without losing excessive moisture) and use the carbon dioxide during the day.

    While the PEPCO using plants are not as efficient at ordinary temperatures as plants employing Rubisco, once the temperature begins the rise above about 30 degrees Celsius, the productivity of the Rubisco plants falls off. The PEPCO plants, however, remain productive at higher temperatures. Thus the use of PEPCO to avoid the penalty of photorespiration is an advantage under hot conditions, and in late summer in temperate climates or in the tropics this makes up for lower productivity during cooler weather **(C3, C4 and CAM plants)**

53. C_3 plants include many of the most common plants, such as wheat and rice. C_4 plants include sugar cane and corn, as well as many crabgrasses. **(C3, C4 and CAM plants)**

Grade Yourself

Circle the question numbers that you had incorrect. Then indicate the number of questions you missed. If you answered more than three questions incorrectly, you will then have to focus on that topic. (If a topic has less than three questions and you had at least one wrong, we suggest you study that topic also. Read your textbook, a review book, or ask your teacher for help.)

Subject: Photosynthesis

Topic	Question Numbers	Number Incorrect
Light	1, 2, 3, 4, 5, 6, 28	
Chloroplast—structure and function	7, 8, 9	
Overview of photosynthesis	10, 11, 12, 13, 14, 15, 16, 17, 18, 20, 27, 51	
Light-dependent reactions	19, 21, 22, 23, 24, 25, 26, 29, 30, 31, 32, 33, 34, 35, 42	
Light-independent reactions	36, 37, 38, 39, 40, 41, 43, 44, 45, 46, 47, 48	
C3, C4 and CAM plants	49, 50, 52, 53	

Mendelian Genetics

7

Brief Yourself

Basic Concepts and Terms, Mendel's Experiments

Genetics is the study of inheritance—how genes are transmitted from generation to generation. These genes can be described in terms of the characteristics they code for (tallness versus shortness in garden peas) or as sections of DNA on chromosomes in cells which code for particular proteins. Genes may exist in one or more alternative forms called alleles. Gregor Mendel (1822–1884) observed the effects of crossing different strains of the garden pea. In garden peas the gene for pod color has two alleles, *G* for green pod and *g* for yellow. The capital letter stands for the dominant allele and the small for the recessive allele. Genes have a specific "address" or location on a chromosome, called a locus, and the locus is occupied by a particular allele. Genes can be considered particles which are inherited as discrete units, and remain as separate units during transmission from parent to offspring. This is in contrast to the 19th-century theory of blending inheritance, which stated that parental characters blend to produce an intermediate character in the offspring.

Monohybrid and Dihybrid Genetics

Crosses can be made between pure strains of organisms that differ with respect to one trait—e.g., mating a black guinea pig with a white guinea pig (the *P* generation), symbolized as

BB x *bb*

where *B* stands for the dominant allele for black and *b* for the recessive. This is a monohybrid cross. If the strains differ in a second trait (e.g., long hair, *L*, or short hair, *l*)) the cross is described as dihybrid and written as:

BBLL x *bbll*

for a cross between black, long haired guinea pig and a white, short-haired animal. The offspring are the F_1 (first filial) generation.

Testcrosses

The dominant phenotype can result from a homozygous or heterozygous genotype. To distinguish between the two, a testcross is performed using a homozygous recessive organism, in which the phenotype reflects the specific genotype. The phenotypes found in the F_1 will reveal the genotype of the other parent.

Multiple Alleles

Genes may exist in only one form or in multiple forms as two or more alleles. While an individual diploid organism can contain at most two different alleles, a number of other distinct alleles could exist in other members of that population.

Polygenic Inheritance

Many quantitative characteristics such as height are affected by a number of genes, each making a contribution to the phenotype. The various combinations of alleles and genes produces a range of heights in a population. When the measured heights for members of a population are graphed, the curve (distribution) is bell-shaped.

Test Yourself

1. Alternative forms of a gene (e.g., short and tall in garden peas, *t* and *T*) are called _________.

2. The "address" of a gene on a chromosome is called its ___________.

3. True or false: Mendel's genetics are a form of blending inheritance.

4. Which of the following is not true of Mendel?

   a. Mendel did his work with strains of garden peas.

   b. Mendel was born and did his work in the early 20th century.

   c. Mendel's work was forgotten for nearly forty years and only rediscovered after his death.

   d. Gregor Mendel was a monk and teacher of mathematics and physics.

5. Distinguish between monohybrid and dihybrid.

6. In instances of true dominance-recessiveness, why are the homozygous recessive organisms valuable for investigating genetics?

7. What types of gametes, and in what proportions can be derived from parents of the following genotypes? (i) *Gg*, (ii) *TT*, (iii) *tt*

8. Label the following as (i) diploid or haploid genotype, and (ii) as homozygous dominant, homozygous recessive or heterozygous, dominant or recessive.

   a. *GG*

   b. *Ss*

   c. *t*

   d. *rr*

9. What are the genotypes, phenotypes, and ratios of the F1 generations from the following crosses: (I) *TT* x *tt*, (ii) *TT* x *Tt*,, (iii) *tt* x *Tt*, (iv) *Tt* x *Tt*.

10. Crossing a red homozygous dominant snapdragon with a white homozygous recessive snapdragon yields F1 plants with white, pink and red flowers in a ratio of 1:2:1. How can this case be explained as an example of particulate inheritance rather than blending inheritance, when a cross of red and white plants gives some pink offspring?

11. Give an example of each of the following types of interactions between alleles:

    (i) True dominance—recessiveness

    (ii) Incomplete dominance

    (iii) Codominance

12. Briefly explain the differences between the three types of allelic interaction list in question 11.

13. How can there be multiple alleles (more than two) for a gene if a diploid organism can have at most two different alleles in its genotype?

14. List the human blood types (phenotypes) and their genotypes according to the *ABO* system discovered by Landsteiner.

15. Distinguish between phenotype and genotype, giving an example of each.

16. In your own words, state Mendel's Law of Segregation (Mendel's First Law) and what it says about alleles, segregation, and gametes.

17. In your own words, state Mendel's Law of Independent Assortment (Mendel's Second Law).

18. What is the abbreviation for (i) first filial generation; (ii) second filial generation.

19. The offspring resulting from a cross between two pure homozygous recessives would be:
    a. 50 percent homozygous recessive, 50 percent heterozygous
    b. 25 percent homozygous recessive, 75 percent heterozygous
    c. 50 percent homozygous recessive, 50 percent homozygous dominant
    d. 100 percent homozygous recessive

20. In a certain species of brown mice, brown fur is dominant over white fur and long tails are dominant over short tails. Both of these traits are inherited independently of each other. With respect to only these traits, how many different phenotypes would be present in a large population of mice?
    a. 1
    b. 2
    c. 3
    d. 4
    e. 5

21. In garden peas, plants with garden green pods have one of two genotypes, *GG* or *Gg*. How can you tell which genotype is present in a particular plant?

22. In a desert rat, black hair (*B*) is dominant over white hair (*b*). A black-haired female and white-haired male are crossed, and the resulting litter consists of four black and one white desert rats. The genotype of the mother is:
    a. *B*/*B*
    b. *b*/*b*
    c. *B*/*b*
    d. can't be determined from this information

23. What is meant by *penetrance*? Give an example of a gene that shows 100 percent penetrance in humans.

24. A gene that affects the phenotype in multiple ways is termed ___________.

25. What is epistasis? Give an example of epistasis.

26. Human skin color is determined by three or more genes. The presence of dominant alleles for each gene produces the darkest skin color, and the presence of recessive alleles for each gene yields the lightest skin color. The shades in between are produced by various combinations of recessive and dominant alleles. This is a classic example of ___________ inheritance.

27. Refer to question 26. Draw a simple graph showing the distribution of shades of skin color in a large, randomly selected group of people.

28. Fill in the following table, indicating the possible blood types of the children from parents of the listed phenotype (blood type). A completed example heads the list.

Parental blood types	Possible blood types of children
A x A	A, O
A x B	
A x O	
B x O	
AB x AB	
AB x O	
AB x A	

29. A couple with blood types A and O has three children, all blood type O. What are the chances that a fourth child will have blood type O?

30. A man with blood type AB has children with a woman of blood type AB. What are the possible genotypes of their children?

31. What are the possible genotypes of sperm or eggs produced by a pea plant with a diploid genotype of *Tt GG Ii* ?

32. In a desert rat, black hair (*B*) is dominant over white hair (*b*). A black-haired female and white-haired male are crossed, and the resulting litter consists of five black desert rats. The genotype of the mother is

    a. *B*/*B*

    b. *b*/*b*

    c. *B*/*b*

    d. can't be determined from this information

Fill in the blank for questions 33–38.

Possible answers for 33–38: a. morphs,
b. recombinants, c. pleiotrophy, d. epistasis, e. locus, f. address, g. combination, h. linkage group, i. alleles, j. incomplete dominance,
k. codominance

33. Where a gene is found____________

34. One gene interferes with or masks the effect of another gene____________

35. Genes that tend to stay together on a chromosome during meiosis____________

36. Gene has multiple effects on phenotype as in Marfans syndrome____________

37. Genes for human blood types show this____________

38. Genes for flower color in snapdragons and petunias are examples of this___________

39. In cattle, the genotype *RR* is for red coat. The allele *R* for red color is not fully dominant over the allele for white coat color, *r*. The heterozygote is an intermediate type called roan, with a dark coat sprinkled with white or gray.

    If a roan cow is crossed with a roan bull, list the following for the F_1 offspring:

    genotypes ____________________
    ratios ____________________
    phenotypes (coat color)____________________
    ratios ____________________

40. In peas, gray seed color (*G*) is dominant over green (*g*). A series of crosses where performed and the information listed below collected. Determine the genotype of each parent.

Parental phenotypes	Parental genotypes	F1 phenotypes and numbers	
		Gray	Green
a. Gray x Gray	_______	62	19
b. Gray x Gray	_______	87	0
c. Green x Green	_______	0	127
d. Gray x Green	_______	65	59

41. Give the phenotypic and genotypic results (including ratios) for this dihybrid cross between two guinea pigs. The genes are found on two different chromosomes and exhibit complete dominance-recessiveness.

    coat color: *B* (dominant, black), *b* (recessive, white)

    hair length: *L* (dominant, long), *l* (recessive, short)

    BbLl x *bbll*

42. What specific type of cross is used in question 41?

Check Yourself

1. alleles **(Blending versus particulate inheritance)**

2. locus **(Blending versus particulate inheritance)**

3. False. Mendelian genetics involve particulate inheritance, where the genes are the particles. An example will clarify the difference between these models. If a true-breeding, white-flower petunia is crossed with a red-flowered petunia from a pure strain, a pink-flowered plant results. Explained as blending inheritance, the pink color comes from a mixing of the genes for red and white. The particulate theory explains that while the mixing of the flower pigments gives a pink color, the genes themselves are not mixed or blended. The correctness of the latter explanation may be proven by crossing two pink-flowered plants. Plants with red, white and pink flowers are obtained, showing the original, discrete red and white genes are still present in the pink-flowered parents. **(Blending versus particulate inheritance)**

4. b. Mendel lived his entire life during the 19th century. **(Basic concepts and terms, Mendel's experiments)**

5. A hybrid strain of plant or animal results when two different strains are crossed. In a monohybrid cross, the strain differs with respect to one inherited characteristic, in a dihybrid cross the strains differ in two characteristics. **(Monohybrid and dihybrid genetics)**

6. The phenotype and genotype are only specifically known for the homozygous (doubly) recessive—short pea plants have a genotype of *tt*, whereas tall pea plants may be either *TT* or *Tt*. The underlying differences between the tall and short can be seen by comparing the genotype (genetic makeup) and phenotype (physical body) of a pea plant. If the plant has a diploid genotype of *GG* (called homozygous dominant) the dominant allele is expressed and the pod color is green. The dominant allele is also expressed in the instance of the mixed *Gg* genotype (heterozygote). The recessive allele is expressed in the phenotype on when it is paired with another recessive allele, as in the genotype *gg*. (Note that except in cases such as human *ABO* blood types, the phenotype is never simply an expression of the genotype—it is a result of the combined effects of the genotype and the environment.) **(Monohybrid and dihybrid genetics)**

7. (i) One-half the gametes will bear the *G* allele, and the other half the *g* allele.

   (ii) All gametes are *T*.

   (iii) All gametes are *t*. **(Blending versus particulate inheritance)**

8. a. diploid genotype, homozygous dominant

   b. diploid, heterozygous

   c. haploid, recessive

   d. diploid, homozygous recessive **(Monohybrid and dihybrid genetics)**

9. (i) All *Tt*

   (ii) *TT* : *Tt*, 1:1

   (iii) *tt* : *Tt*, 1:1

   (iv) *TT* : *Tt* : *tt*, 1:2:1 **(Monohybrid and dihybrid genetics)**

10. The pink flower color is due to the weakening or dilution of the red color by the white—the colors mix to give pink, but the genes themselves do not blend, and the genes retain their discrete identities. This can be shown by crossing two pink snap dragons—the resulting offspring will be red : pink : white in a ratio of 1:2:1. Pink flowers bear both the red and white alleles. **(Degrees of dominance)**

11. (i) tall versus short in pea plants

    (ii) flower color in snap dragons (see answer to question 10)

    (iii) human blood types

    Alleles can interact in a number of ways, in addition to showing a simple dominant-recessive relationship as in the examples given above of garden peas and guinea pigs, alleles may also be codominant (*ABO* blood types) or incompletely dominant (petunia flower color). **(Interactions between genes)**

12. (i) If the dominant allele is present it is expressed, while the recessive allele is expressed only in the homozygous recessive state.

    (ii) The alleles are either dominate or recessive. Both alleles are observable in the heterozygote.

    (iii) Two alleles are equally expressed—for instance, in the human blood type *AB* both of the alleles are expressed. **(Degrees of dominance)**

13. While only two alleles of a gene may exist in the body of an individual diploid organism, other individuals in that population or species may possess other alleles. **(Multiple alleles)**

14.

blood types	genotypes
A	*AA, AO*
B	*BB, BO*
AB	*AB*
O	*OO* **(Degrees of dominance)**

15. Genotype refers to the genetic information contained in the alleles on the chromosomes in an organism. Phenotype is a characteristic you can see—usually the result of the interplay between heredity and the environment. (In the case of ABO blood types, however, the phenotype is entirely determined by heredity and is not influenced by the environment.) **(Basic concepts and terms, Mendel's experiments)**

16. Stated as we understand it today, Mendel's First Law states that alleles remain discrete, even in the heterozygote state; that alleles segregate during meiosis; and that chance determines which allele will go to a particular gamete. **(Basic concepts and terms, Mendel's experiments)**

17. Mendel's Second Law says that genes on different chromosomes assort independently during meiosis. **(Basic concepts and terms, Mendel's experiments)**

18. (i) F_1 (ii) F_2 **(Monohybrid and dihybrid genetics)**

19. d. For instance, parents in the cross *rr* x *rr* can only pass *r* alleles on to their offspring. **(Monohybrid and dihybrid genetics)**

20. d. 4 **(Monohybrid and dihybrid genetics)**

21. The homozygous dominant form may be distinguished from the heterozygote by using a testcross—crossing the plant in question with a homozygous recessive plant. **(Testcrosses)**

22. c. B/b **(Testcrosses)**

23. Penetrance refers to the proportion of individuals bearing a gene who express it phenotypically. In the case of the ABO blood typing system, Penetrance is 100 percent—the phenotype completely reflects the genotype. Incomplete Penetrance can be seen in hereditary diseases, where the disorder is expressed to different degrees of severity in different individuals. **(Degrees of dominance)**

24. pleiotrophic (the noun is pleiotrophy) **(Interactions between genes)**

25. Consider two pairs of non-allelic genes, such as in peas where a dominate allele (*P*) produces purple flowers, and another gene affects the series of chemical reactions (metabolic pathway) which produces the pigment for flower color. If the pigment production gene is homozygous recessive, the flower will be white, even if the genotype of the plant is *PP* or *Pp*. The alleles of the second gene "override" the alleles of the first. **(Interactions between genes)**

26. polygenic **(Polygenic inheritance)**

27. The graph should show a distribution of skin shades that is bell-shaped. Relatively fewer people will have the darkest or lightest skin colors, with the majority of people having an intermediate skin color. **(Polygenic inheritance)**

28. Solve problems of this type by determining the possible genotypes of the parents. In the example given the parents both have blood type A but their genotypes may be *AA* or *AO*. Hence the children may have blood type O (genotypes *OO*) or blood type A (genotypes *AA, AO)*

Parental blood types	Possible blood types of children
A x A	A, O
A x B	A, B., AB, O
A x O	A, O
B x O	B, O
AB x AB	A, BA, B, AB (Note BA = AB!)
AB x O	A, B
AB x A	A, B, AB **(Degrees of dominance)**

29. Each child has a 50-50 chance of having blood type O. Simply because three children have blood type O doesn't change the odds for the fourth child. **(Monohybrid and dihybrid genetics)**

30. AA, BB, and AB **(Degrees of dominance)**

31. *TGI, TGi, tGI, tGi* **(Monohybrid and dihybrid genetics)**

32. d. The small size of the sample makes it impossible to be certain of the genotype of the mother. Was the mother a heterozygote and merely due to chance that all the rats in the litter were black, or was the mother homozygous dominant? The question cannot be answered with information provided. **(Testcrosses)**

33. e. locus **(Basic concepts and terms, Mendel's experiments)**

34. d. epistasis **(Interactions between genes)**

35. h. linkage group **(Basic concepts and terms, Mendel's experiments)**

36. c. pleiotrophy **(Basic concepts and terms, Mendel's experiments)**

37. k. codominance **(Degrees of dominance)**

38. j. Different (non-allelic) genes can also interact. Epistasis describes a case in which gene at one locus prevents the expression of a gene at another locus. (Pleiotrophy, which is frequently confused with epistasis, refers to a gene having multiple effects on the phenotype. For instance the gene for Marfan's syndrome affects connective tissue, and therefore numerous parts of the human body.) **(Degrees of dominance)**

39. genotypes *RR : Rr : rr*

    ratios 1:2:1

    phenotypes (coat color) red : roan : white

    ratios 1:2:1

    This problem is analogous to those concerning flower color in snap dragons and petunias. **(Degrees of dominance)**

40.

Parental phenotypes	Parental genotypes	F1 phenotypes and numbers		
		Gray	Green	(ratio)
a. Gray x Gray	*Gg*	*62*	*19*	*(3:1)*
b. Gray x Gray	*GG, Gg*	*87*	*0*	*(100%)*
c. Green x green	*gg*	0	127	(100%)
d. Gray x green	*GG, gg*	*65*	*59*	*(1:1)* **(Monohybrid and dihybrid genetics)**

41.

genotype	ratio	phenotype
BbLl	25%	black, long
Bbll	25%	black, short
bbLl	25%	white, long
bbll	25%	white, short **(Monohybrid and dihybrid genetics)**

42. A testcross **(Testcrosses)**

Grade Yourself

Circle the question numbers that you had incorrect. Then indicate the number of questions you missed. If you answered more than three questions incorrectly, you will then have to focus on that topic. (If a topic has less than three questions and you had at least one wrong, we suggest you study that topic also. Read your textbook, a review book, or ask your teacher for help.)

Subject: Mendelian Genetics

Topic	Question Numbers	Number Incorrect
Blending versus particulate inheritance	1, 2, 3, 7	
Basic concepts and terms, Mendel's experiments	4, 15, 16, 17, 33, 35, 36	
Monohybrid and dihybrid genetics	5, 6, 8, 9, 18, 19, 20, 29, 31, 40, 41	
Degrees of dominance	10, 12, 14, 23, 28, 30, 37, 38, 39	
Interactions between genes	11, 24, 25, 34	
Multiple alleles	13	
Testcrosses	21, 22, 32, 42	
Polygenic inheritance	26, 27	

Mitosis, Meiosis, Chromosomes, Genes, and DNA

Brief Yourself

DNA, Chromatin and Eukaryotic Chromosomes

Chromatin is composed of DNA complexed with proteins, the histones and the nonhistone chromosomal proteins. Chromosomes are chromatin tightly coiled into visible bodies. When relaxed, they form long, thin threadlike molecules.

Eukaryotic Cell Cycle

The eukaryotic cell cycle may be broken into several parts: growth, chromosomal replication, mitosis, cytokinesis. Not all cells progress through the entire cycle. Nerve cells of higher vertebrates, for instance, do not normally divide in the adult animal. In the G_1 stage, the new daughter cells resulting from mitosis and cytokinesis grow into normal-sized cells. Replication of DNA occurs next, in the S stage. In the G_2 stage, cell organelle numbers are doubled in preparation for cell division.

Overview of Mitosis and Meiosis

During mitosis, two new daughter cells are produced (actually two new daughter nuclei) which have the same complement of chromosomes as was found in the parent cell. Whether the parent cell was haploid or diploid, the daughter cells will retain that number of chromosomes. Meiosis is a case of reduction division, where a diploid is reduced to a haploid; a parent diploid germ-line cell giving rise to haploid gametes.

Haploid and Diploid

Chromosomes in eukaryotes are found in sets. A single complete set is the haploid number of chromosomes. Two haploid sets (one derived from the organism's mother, the other from the organism's father) make up a diploid set. Human beings have a haploid (1N) number of 23 chromosomes, and a diploid (2N) number of 46 chromosomes (or 23 pairs). Additional sets of chromosomes may be found in instances of polyploidy, such as triploidy (three sets, or 3N). In plants, the multicellular haploid stage of the life cycle is called the gametophyte, and the diploid stage is the sporophyte.

Meiosis

Meiosis is divided into two general phases, meiosis I and meiosis II. In each phase, the stages bear the same name as in analogous stages of mitosis. In prophase I (prophase of meiosis I), chromosomes cross over during synapsis at sites termed chiasmata. In metaphase I, chromosomes independently assort, followed by separation of homologous chromosomes in anaphase I. During meiosis II sister chromatids separate.

Test Yourself

1. On what cellular structure are genes carried in eukaryotes? Do prokaryotes such as bacteria have the same sort of structures?

2. Imagine you have two lumps of clay, one red and one green. Roll each one into a cylinder (like a hot dog) and lay them along side each other. Near the middle of each cylinder, pull the clay together, pinching them so the rolls now touch. (The rolls now form an x-shape.) Each roll represents a chromosome. Keeping your model in mind, answer these questions.

   (i) In this model, what represented one chromosome before the rolls of clay were pinched together?

   (ii) After pinching the rolls together, what represents one chromosome?

   (iii) What feature would be located near the middle where the two rolls join (the word central is a hint).

   (iv) What part of the model is a chromatid? In other words, would a chromatid on this model be one-half red and one-half green, or all one color?

3. Read the following three statements. One describes eukaryotic chromosomes, another describes prokaryotic chromosomes, and the remaining one is an incorrect statement. Label the statements appropriately as "eukaryotic," "prokaryotic," or "false."

   (i) Composed of condensed DNA and proteins such as histones; DNA is coiled about some histones, forming nucleosomes.

   (ii) The DNA is in the form of a circle and complexed with many proteins, including histones.

   (iii) The DNA is in the form of a circle and complexed with only a few proteins.

4. Do prokaryotes such as bacteria have chromosomes?

Questions 5–13 Matching.

Possible answers:

a. The process of the cytoplasm a cell dividing in two.

b. A single complement or set of chromosomes.

c. Nuclear division, maintaining the same number of chromosomes in the daughter nuclei as in the parent nucleus.

d. Nuclear reduction division, in which the number of chromosomes per cell is cut in half.

e. Duplicate chromosomes joined in the region of the centromere will separate to become daughter chromosomes

f. Another name for body cells.

g. Another name for the cell line involved in the production of gametes.

h. In bacteria, the process of replicating the genetic material and then dividing into two daughter cells.

i. A double complement (set) of chromosomes.

5. Sister chromatid

6. Haploid

7. Diploid

8. Cytokinesis

9. Mitosis

10. Meiosis

11. Binary fission

12. Germ cells

13. Somatic cells

14. How many chromosomes are there in:

    (i) Human somatic cells?

    (ii) Human gametes?

15. A person has two sets of chromosomes in all somatic cells. Where did each original set come from?

16. Humans have _____ pair(s) of autosomes and ___ pair(s) of sex chromosomes. The male karyotype for the sex chromosomes is ____, the female karyotype is ____.

17. What moves the chromosomes during cell division?

18. What are the stages of mitosis? What happens to the DNA or chromosomes during each stage? During which stages is the nucleolus visible?

19. What happens to chromosome numbers during mitosis? Can a haploid cell undergo mitosis, and if it does, what type of cell (diploid or haploid) result?

20. When in the cell cycle is the DNA replicated for mitosis?

21. Do all cells undergo mitosis? Give examples of cells that usually do and don't.

22. What is accomplished in (Meiosis) meiosis I and (Meiosis) meiosis II?

23. When in meiosis do the chromosomes synapse?

24. When in meiosis does crossing over occur?

25. When in meiosis does independent segregation of chromosomes occur?

26. What are homologous chromosomes?

27. Dogs have a 2N number of 78.

    (i) Dog sperm have how many chromosomes?

    (ii) Diploid dog cells have how many chromosomes?

    (iii) How many chromosomes would you find in a dog muscle cell?

    (iv) What is the haploid number of chromosomes in a dog?

28. Sometimes homologous chromosomes fail to separate during Meiosis I or chromatids fail to separate in Meiosis II. This abnormality is called __________.

29. Genes that tend to stay together on a single chromosome during meiosis are called _______ __________. (two words).

30. Morgan's work with fruit flies, in which he followed the inheritance of eye color (red and white) in males and females, showed that genes for some traits were on the sex chromosomes—these traits are called_______ - ___________ traits. (two words)

31. During what part of the cell cycle are the organelles duplicated?

32. What is the distinction between mitosis and cytokinesis?

33. Pick examples of organisms which use each of the following systems of sex determination:

    (i) X-Y

    (ii) X-O

    (iii) Z-W

Possible choices: humans and most other mammals; grasshoppers and roaches; birds and butterflies.

34. The period of cell growth following cell division is the ____ stage of the cell cycle.

    a. G_1

    b. G_2

    c. S

    d. mitosis

35. Describe a difference between plant and animal cells during cell reproduction.

36. Distinguish between sexual and asexual reproduction.

37. Where do the sperm and eggs originate in an angiosperm (flowering plants)?

38. At what point in the reproductive cycle is the parent number of chromosomes restored?

39. True or false; if false correct to make the statement true: The sporophyte in plants develops from the zygote by mitosis and produces spores by meiosis; the gametophyte develops from spores and produces gametes.

40. Indicate as being haploid or diploid:

    a. sporophyte

    b. gametophyte

    c. gametes

    d. somatic animal cells

41. What are several common differences between eggs and sperm?

42. The process of meiosis results in ___ sperm, but only ___ egg(s) and ___ polar bodies.

Questions 43–45. Match the name with the description of the chromosome change.

(i) The loss of a section of a chromosome.

(ii) A section of chromosome is removed, and replaced in the reverse order.

(iii) Repetition of gene sequences.

43. Deletion

44. Inversion

45. Duplication

Possible answers:

46. What is aneuploidy?

47. What is polyploidy? Is polyploidy common or rare in plants? In animals? Give an example.

48. How are chromosomes mapped—which of the following are true for mapping chromosomes?

    (i) Experimental crosses are performed and the percentage of recombinants gives relative distances between genes.

    (ii) Distances between genes are stated in terms of map units (also called centimorgans).

    (iii) Under the transmission electron microscope, the distance between genes is measured.

    (iv) Genes that are closer are more likely to stay together (as a linkage group) than genes that are farther apart, and this fact underlies all mapping.

49. True or false; If false correct to make the statement true. Males are genotype XX and females are genotype XY with respect to their autosomes.

50. The first people to map chromosomes were _______ and co-workers.

    (i) Mendel

    (ii) Meselson

    (iii) Morgan

    (iv) Mitchell

51. When sex-linked traits are referred to, one should always think of the ____ (X or Y) chromosome.

52. What is the end effect of crossing over, independent segregation of chromosomes, and sexual reproduction (meaning syngamy, or fusion of gametes)?

53. What is a recombinant?

Check Yourself

1. Genes are carried on chromosomes in eukaryotes. Bacteria have what can be called a form of a chromosome, different from the true chromosome of eukaryotes in that fewer proteins (and no histones) are involved, and the bacterial DNA exists in the form of a loop. Reproduction is most frequently caused by binary fission, an asexual process in which the DNA is duplicated and two new daughter cells produced. Prokaryotes are also capable of sexual reproduction. Some textbooks and instructors avoid referring to "bacterial chromosomes." **(Prokaryotic cell division)**

2. (i) Each cylinder of clay represents one chromosome, as stated.

   (ii) Each cylinder of clay now represents one chromatid (sister chromatid); the two together are called a chromosome.

   (iii) Centromere, or central (centro-) unit (-mer, as in polymer).

   (iv) One chromatid would be green, the other would be red. The chromosome is divided lengthwise into sister chromatids. **(Structure of chromosomes)**

3. (i) eukaryotic

   (ii) false

   (iii) prokaryotic

   Eukaryotic chromosomes are highly organized structures in which a strand of DNA forms a complex with proteins, most notably histones. At intervals the DNA is wound twice around a group of eight histone molecules, forming a nucleosome. The DNA is looped and coiled further, forming the highly condensed chromosome. **(Structure of chromosomes)**

4. Yes and no. The bacterial chromosomes are distinctly different from eukaryotic chromosomes, different enough to cause some teachers to avoid applying the word "chromosome" to bacteria. See answer to question 1, above. **(Prokaryotic cell division)**

5. e **(Structure of chromosomes)**

6. b **(Haploid and diploid)**

7. i **(Haploid and diploid)**

8. a **(Mitosis and cytokinesis)**

9. c. Mitosis, strictly speaking, is nuclear division. Cytokinesis is the physical division of the parent cell into two daughter cells. Mitosis can occur with or without subsequent cytokinesis. In mitosis, the chromosomes double in number and two new daughter nuclei form. Muscle cells frequently undergo mitosis without cytokinesis, and as a result have multiple nuclei. In cytokinesis, a cell or cleavage furrow (animals cells) or cell plate (plant cells) forms across the cell at right angles to the direction of chromosome movement, and the cell divides into two smaller, daughter cells. **(Mitosis and cytokinesis)**

10. d **(Meiosis)**

11. h **(Prokaryotic cell division)**

12. g. **(Haploid and diploid)**

13. f **(Haploid and diploid)**

14. (i) 46 individual chromosomes, or 23 pairs

    (ii) 23 individual chromosomes **(Sexual and asexual reproduction)**

15. One haploid set came form the person's mother, the other haploid set from the person's father. **(Haploid and diploid)**

16. 22, 1, XY, XX **(Sex-determination; linkage)**

17. microtubules powered by ATP **(Mitosis and cytokinesis)**

18. The stages of mitosis and associated characteristics are:

    (i) prophase, in which the chromosomes condense and become visible, and the nucleolus and nuclear envelope disappear, and the mitotic spindle appears.

    (ii) metaphase, in which the chromosome line up at the equator of the cell spindle, along the line where the cell will later divide in two.

    (iii) anaphase—the chromosomes come apart and are moved by microtubules toward opposite ends of the cell, often assuming an inverted V-shaped.

    (iv) telophase, which is actually the reverse of prophase—the nuclear envelopes reform around the daughter chromosomes. The chromosomes condense, the nuclear envelope reassembles, the nucleoli appear.

    Cytokinesis is not part of mitosis, but often concludes it. It is the actual process of dividing the cytoplasm of the parent cell into two daughter cells. **(Mitosis and cytokinesis)**

19. Chromosome numbers stay constant during mitosis—a cell with 4 pairs of chromosomes (as in *Drosophila*) will divide into two cells, each with 4 pairs of chromosomes. A haploid cell can undergo mitosis and the result will be two haploid daughter cells. The rule still applies, haploid or diploid: mitosis keeps the chromosome numbers the same. **(Mitosis and cytokinesis)**

20. During the S phase **(Cell cycle)**

21. No, not all cells undergo mitosis. In humans, skins cells are constantly dividing to replace lost cells, but muscle cells rarely divide in adults. Liver cells don't normally divide, but will do so if part of the liver is removed. **(Mitosis and cytokinesis)**

22. (i) In meiosis I, crossing over occurs and homologous chromosomes are separated.

    (ii) During meiosis II, sister chromatids are separated. **(Meiosis)**

23. In prophase I, chromosomes synapse and crossing over occurs. **(Meiosis)**

24. In prophase I, chromosomes synapse and crossing over occurs. **(Meiosis)**

25. The actual separation occurs in anaphase I. **(Meiosis)**

26. Homologous chromosome are pairs of chromosomes, one chromosome from the mother and the other from the father of the organism. Pairs of chromosomes in diploid sets exist as homologous chromosomes. There are two chromosomes number one, two chromosomes number two and so on in a diploid organism. Homologous chromosomes contain the same genes but differ in the specific alleles found at the various loci (gene addresses on the chromosome). **(Meiosis)**

27. (i) 39

    (ii) 78

    (iii) 78 or 39 pairs

    (iv) 39 **(Haploid and diploid)**

28. nondisjunction, a very descriptive term **(Changes in chromosome number and structure)**

29. closely linked. Genes that are located more closely together on a chromosome are more likely to remain together (i.e., avoid separation by crossing over) over time than genes that are more dispersed. Such genes move through generations as a unit and are termed closely-linked. Genes on a chromosome are said to make up a linkage group. **(Sex-determination; linkage)**

30. Sex-linked (that is, located on the X chromosome), not to be confused with sex-associated traits which are simply associated with one or the other sex traits. **(Sex-determination; linkage)**

31. G_2. **(Cell cycle)**

32. Mitosis is nuclear division which conserves the number of chromosomes. Cytokinesis is the division of the cytoplasm of one cell into two daughter cells. Not all cells that undergo mitosis follow nuclear divisoin with cytokinesis. Muscle cells commonly have multiple nuclei as a result of undergoing mitosis without subsequent cytokinesis. **(Mitosis and cytokinesis)**

33. (i) humans and most other mammals (and fruit flies as well)

    (ii) grasshoppers and roaches

    (iii) birds, butterflies (and moths)

    The sex of an organism can be determined by one of several systems. In the X-Y system, found in humans and most mammals, the male is XY (heterogametic) and the female is XX. In the X-O system found in some insects such as grasshoppers, males possess only a single X chromosome, and their karyotype is designated XO. Females have two X chromosomes, or XX. In the third system Z-W, found in birds and fish, females are ZW (one Z chromosome and one W chromosome) and males are ZZ. **(Sexual and asexual reproduction)**

34. a. G_1 **(Cell cycle)**

35. Animal cell division differs from plant cell division in two respects: (i) animal cells form cleavage furrows and plant cells form cell plates, due to plants having cell walls; and (ii) animal cells have centrioles and asters (made of microtubules). **(Sexual and asexual reproduction)**

36. In asexual reproduction, an organism duplicates itself without genotypic variation. For example, a protist like an *Amoeba* divides in two, or a bacterium splits into two daughter cells (binary fission), or a *Hydra* or yeast produces a bud. Parent and offspring are genetically alike. In sexual reproduction, haploid cells (gametes) from different parents come together (a process called fertilization or syngamy) to form a new, diploid individual, genetically distinct from either parent. Asexual reproduction tends to be advantageous in stable environments, whereas sexual reproduction, by offering novel combinations of genes from two parents, can be an advantage in changing environments. **(Sexual and asexual reproduction)**

37. Sperm come from the anthers, while the eggs are produced and develop in the ovary. **(Meiosis)**

38. This occurs during sexual reproduction. During syngamy, or fertilization, when the nucleus of the sperm fuses with the nucleus of the egg. **(Sexual and asexual reproduction)**

39. True (**Meiosis**)

40. a. The sporophyte is diploid.

    b. The gametophyte is haploid.

    c. Gametes are haploid.

    d. Somatic animal cells are diploid. (**Haploid and diploid**)

41. Sperm are usually smaller than eggs, and motile. Eggs are usually larger than sperm, contain foodstuffs for the new organism, and are sessile. (**Meiosis**)

42. 4, 1, 3 (**Meiosis**)

43. i (**Changes in chromosome number and structure**)

44. ii (**Changes in chromosome number and structure**)

45. iii. If homologous chromosomes or sister chromatids do not properly separate during meosis, a condition known as nondisjucntion results, in which one haploid daughter gamete gets two copies of a specifc chromosome, and another gamete recieves no copies. This outcome is termed aneuploidy. Sometimes multiple sets of chromosomes greater than the diploid number accumualte in a cell, a state of polyploidy. While rarely observed in animals, polyploidy is common in many plants and has played a significant role in the evolution of plants. In crossing over, mistakes can be made in reassembling chromosomes—sections can be omitted (deletion), duplicated, or reversed in order (inversion). Many of these mistakes can have important or profound consequences. (**Changes in chromosome number and structure**)

46. Aneuploidy results from nondisjunction—during meiosis, one gamete receives both copies of a chromosome and another gamete receives no copy. Zygotes that result from fertilization with one of these gametes have an abnormal number of chromosomes. Down's syndrome in people is an example of aneuploidy, specifically a trisomy (having three) chromosome number 21. (**Changes in chromosome number and structure**)

47. Polyploidy means there are more than the diploid number of sets of chromosomes—diploid is two sets, triploid is three, and tetraploid means four sets. Polyploidy is common in plants, such as angiosperms (flowering plants) and has undoubtedly played a significant role in the evolution of plants such as ferns. It is rare in animals. (**Changes in chromosome number and structure**)

48. Answers i, ii, iv are true; iii is false. A relative map of chromosomes—where certain genes are located with respect to other genes—can be established by careful experimental crosses. Individuals with unique combinations of traits not seen in their parents (recombinants) allow the construction of a map based on the frequency of genetic recombination. The maps are in map units or centimorgans (after Thomas Hunt Morgan, who pioneered chromosome mapping). (**Sex-determination; linkage**)

49. False. The correct statement is: Males are genotype XY and females are genotype XX with respect to their sex chromosomes. (**Sex-determination; linkage**)

50. iii (**Sex-determination; linkage**)

51. X (**Sex-determination; linkage**)

52. The end result is a new organism, composed of a unique combination of genetic elements, the individual variation that is the raw material for the mill of natural selection. **(Mitosis and cytokinesis)**

53. A recombinant individual has gene combinations not seen in either parent. **(Changes in chromosome number and structure)**

Grade Yourself

Circle the question numbers that you had incorrect. Then indicate the number of questions you missed. If you answered more than three questions incorrectly, you will then have to focus on that topic. (If a topic has less than three questions and you had at least one wrong, we suggest you study that topic also. Read your textbook, a review book, or ask your teacher for help.)

Subject: Mitosis, Meiosis, Chromosomes, Genes, and DNA

Topic	Question Numbers	Number Incorrect
Prokaryotic cell division	1, 4, 11	
Structure of chromosomes	2, 3, 5	
Haploid and diploid	6, 7, 12, 13, 15, 27, 40	
Mitosis and cytokenesis	8, 9, 17, 18, 19, 21, 32, 52	
Meiosis	10, 22, 23, 24, 25, 26, 37, 39, 41, 42	
Sexual and asexual reproduction	14, 33, 35, 36, 38	
Cell cycle	20, 31, 34	
Sex-determination; linkage	16, 29, 30, 48, 49, 50, 51	
Changes in chromosome number and structure	28, 43, 44, 45, 46, 47, 53	

DNA Structure and Replication

Brief Yourself

DNA as the Genetic Material

Deoxyribonucleic acid is the molecule which carries genetic information in virtually all organisms. (Some viruses, however, have double-stranded RNA as their genetic material.) The role of DNA in heredity was established in the 20th century by a number of investigators. Frederick Griffith (1928), in his work on mice using pneumococcus bacteria, discovered that a rough non-pathogenic (doesn't cause disease) strain of bacteria could be turned into a smooth pathogenic strain by the addition of live or killed bacteria from the smooth strain. (The rough or smooth refers to the appearance of the bacterial colonies on agar in petri dishes.) Something, which Griffith called the transforming substance, was being passed from the living or dead S-strain bacteria to the living r-strain bacteria. Griffith's work laid the foundation for the idea that DNA was the genetic material. In 1944 Avery, McLeod, and McCarty published a paper identifying the transforming substance as DNA. In the 1950s, Heresy and Chase showed it was the DNA inside bacteriophage (a virus that attacks bacteria, also called "phage") that was injected into the bacterial cell to make more virus, and that the protein coat remained outside of the cell. In 1953, based on evidence from several sources, Watson and Crick determined the structure of DNA, its phosphate-sugar backbone, specific pairing of purines and pyrimidines, and the meaning of the intramolecular distances. Their model forms the basis for the central idea held concerning DNA. Meselson and Stahl later ascertained that the mode of DNA replication was semi-conservative; that is, the double-stranded daughter DNA molecules were made up of one new strand of DNA and one parent strand.

DNA Structure

DNA consists of long chains of nucleotides strung together to form strands. Two strands are held together in a double helix by hydrogen bonding between the bases adenine and thymine, and the bases guanine and cytosine. This is called complementary base-pairing. Within a strand, the nucleotides are covalently linked to one another in a pattern of sugar - phosphate - sugar - phosphate, with the bases projecting inside the double helix. DNA is long, thin molecule (2 nanometers wide), and the helix makes a turn every 10 nucleotides.

DNA Replication

DNA is replicated by laying down a new strand of deoxyribose nucleotides along a template of parent DNA by the following process. The double-helix is unwound at one point by enzymes which break the hydrogen bonds. The unwinding is stabilized by single stranded binding protein that block the strands from reforming the double helix, and a topoisomerase which prevents the strands from becoming entangled. RNA polymerase attaches a short RNA primer to the DNA parent strand, and

DNA polymerase matches up new nucleotides in a complementary fashion to the parent DNA. DNA polymerase also causes the nucleotides to bond to the new DNA. Replication occurs near the point where the double-helix parts into separate strands, called the replication fork. DNA polymerase removes the RNA primer at the end of replication.

RNA: mRNA, tRNA, and rRNA

Like DNA, ribonucleic acid is made up of nucleotides, but in RNA the sugar is ribose, not deoxyribose, and the base uracil occurs in place of the DNA base thymine.

There are three basic forms of RNA molecules involved in protein synthesis. mRNA carries a sequence of nucleotides that are a transcript of the DNA code. rRNA is found in ribosomes; there are several kinds of rRNA molecules. tRNA is a cross-shaped molecule, with an amino acid on one end and an anticodon on the other. The anticodon consists of three nucleotides which pair up with a complementary sequence of nucleotides on mRNA. This way, a specific amino acid can be added at a given point on a polypeptide chain during protein synthesis.

Test Yourself

1. Griffith's experiment suggested there was a ___________ ___________ (two words) causing the __________ non-pathogenic stain of bacteria, which did not harm mice, to change to the _________ strain which would kill mice.

2. _______________ identified Griffith's substance as DNA.
   a. Avery, MacLeod and McCarty
   b. Watson and Crick
   c. Hershey and Chase
   d. Meselson and Stahl
   e. Beadle and Tatum

3. "Phage" is short for _______________ , a virus that attacks ____________.

4. In the classic "blender experiment," the protein coat of a phage was labeled with a radioisotope of ___________, whereas the DNA inside the virus was labeled with radioactive __________.

5. The experiment mentioned in question 4 was carried out by
   a. Avery, MacLeod, and McCarty
   b. Watson and Crick
   c. Hershey and Chase
   d. Meselson and Stahl
   e. Beadle and Tatum

6. Avery and his coworkers found treated material from the S strain, and found some treatments stopped it from transforming the R strain into the smooth S strain, while other treatments had no effect on the capability of the transforming substance. Of the following procedures, which stopped transformation? Mark these "stopped." Which ones had no effect and did not stop transformation from happening? Label these "no effect."
   a. Treated the transforming substance with DNAse
   b. Treated the transforming substance with RNase
   c. Treated the transforming substance with a protease (protein-digesting enzyme)

7. The double helix model of DNA was proposed in 1954 by
   a. Avery, MacLeod, and McCarty
   b. Watson and Crick
   c. Hershey and Chase
   d. Meselson and Stahl
   e. Beadle and Tatum

8. DNA consists of subunits of ___________. Each subunit is made up of a 5-carbon sugar called __________, a ___________ group, and one of four nitrogen-containing bases, ____________, _____________, ____________, and ____________.

9. Chargaff's rule states that in DNA, the amount (percent) of adenine equals the amount of thymine and the amount of guanine equals the amount of cytosine. If a sample of DNA is found to contain 32 percent adenine, what are the percentages for the other nucleotides?

10. Within a single strand of DNA the phosphate group of one nucleotide is attached to the sugar of the next nucleotide by a __________ bond. In double-stranded DNA, the strands in the helix are held together by __________ bonds between nucleotides on opposite strands.

11. In double-stranded DNA two __________ bonds form between the nucleotides __________ and ___________ on different strands. Three of these bonds form between __________ and __________ on opposite strands.

12. The basis of the structure of DNA is ____________ base pairing.

13. DNA in bacteria is found in the form of _________.

14. During the process of binary fission in prokaryotes, how is half the DNA distributed to each daughter cell?

Questions 15–24. Matching. Answers may be used more than once.

Possible answers:

a. RNA primer
b. DNA polymerase
c. RNA polymerase
d. template
e. helicase
f. topoisomerase
g. single-strand binding proteins
h. Okazaki fragments

15. Unwinds the double helix.

16. RNA needed to start DNA replication.

17. Keeps the two DNA strands from getting tangled.

18. These keep the strands separated during replication.

19. Adds new nucleotides to DNA daughter strands.

20. These are found on the lagging (discontinuous) strand.

21. Proofreads the new DNA strand for mistakes.

22. During DNA replication, each parent strand serves as this.

23. Needed to lay down RNA at the start of replication

24. Removes RNA primer.

25. The mechanism of DNA replication was established as semi-conservative by:

    a. Avery, MacLeod, and McCarty
    b. Watson and Crick
    c. Hershey and Chase
    d. Meselson and Stahl
    e. Beadle and Tatum

26. What is the difference between conservative and semi-conservative DNA replication?

27. Write the DNA nucleotide sequence that is complementary to this length of DNA:

    A T C C G A T C T A A A C G

28. List all products that are made directly or indirectly using DNA as a template.

29. Are the statements made below about DNA polymerase true or false? If false, correct to make the statement true.

    a. DNA polymerase converts nucleotide triphosphates (like ATP, except it has a deoxyribose sugar instead of ribose) into nucleotide monophosphates (like AMP) before placing the new nucleotide onto the DNA template. This process also supplies the energy for adding the nucleotide to the new strand.

b. DNA polymerase occasionally makes mistakes, and attaches the wrong nucleotide to the daughter strand. It corrects this by a process called proofreading in which the wrong nucleotide is cut out and the correct one inserted.

c. DNA polymerase catches all errors in DNA replication.

d. DNA polymerase synthesizes the replicated (daughter) strand in the 3 to 5 direction. This dependence on direction accounts for the formation of leading and lagging strands.

e. DNA polymerase lays down a short length of RNA, called the RNA primer, in order to start replication.

f. Late in replication, RNA polymerase removes the RNA primer from the daughter DNA and replaces it with the proper DNA nucleotides.

30. In what part of the cell does DNA replication occur?

31. Put the steps below in proper order for DNA replication:

a. Nucleotides are covalently bonded together by DNA polymerase to form the new DNA of the daughter strand.

b. The parent DNA strands are unwound and the hydrogen bonds between strands broken. This is accomplished by a helicase enzyme.

c. Complementary nucleotides are added along the parent strand template.

d. RNA polymerase lays down an RNA primer.

32. The standard model of DNA is for the _______ form where the right-handed helix makes a full turn every 10 bases. Another form of DNA, termed ______, consists of a left-handed helix that zig-zags . Possible answers: Z-DNA, B-DNA, A-DNA

33. True or false: All DNA is involved in the synthesis of proteins.

34. Movable genetic units, such as those found by Barbara McClintock in corn or for antibiotic resistance in bacteria, are called __________.

35. Name the four nucleotide bases found in RNA. Which are purines? Which are pyrimidines?

36. List several differences between DNA and RNA.

37. What are the 3 types of RNA molecules?

Questions 38–40. Fill in the blanks.

38. RNA is made on a DNA ___________ through the action of the enzyme ____ ___________ (two words).

39. _______RNA is formed in <u>transcription</u> from DNA, so-called since the genetic message of the DNA is copied (in a complementary fashion) to the_____RNA. (Hint: the same answer is correct for both blanks)

40. ____RNA is involved in _____________ (fill in "translation" or "transcription"), where the genetic message written in RNA nucleotides is changed into the language of amino acids and proteins.

41. A minimum of how many types of tRNA are found in a cell? Why?

42. Roughly sketch a molecule of tRNA, indicating the attachment site for the amino acid and the location of the anticodon.

43. It could be said that complementary base pairing is essential to the production and function of RNA. Explain this statement.

44. Describe two functions of RNA polymerase?

45. Distinguish between a triplet of nucleotides on DNA and a codon. Where are codons found?

46. How many codons are there for the 20 amino acids?

47. A fragment of an mRNA molecule is 180 nucleotides long. Assume all codons are for amino acids, answer these two questions:

a. How many codons does it contain?

b. How many amino acids can it code for?

48. Can only proteins function as enzymes ? Name an exception.

49. Which is larger, a ribosome from a eukaryote or from a prokaryote? Or are they the same size?

50. The ribosome of a bacterium (*Escherichia coli*) is made up of ______ (how many?) subunits, which contain _____ (how many?) different RNA molecules plus numerous proteins. (possible answers: 1, 2, 3, 4)

51. The function of the _________ is to provide a place for the assembly of a polypeptide.

52. For the DNA sequence below, write the complementary mRNA codons and the tRNA anticodons.

    DNA: T C T G A A T G A

    mRNA:

    tRNA

53. The V- shaped points where the double helix comes apart and DNA synthesis occurs are called ______________ ____________ (two words). The entire region where the DNA "bulges" out is referred to as a ____________ bubble.

54. Is the rate of DNA replication (base pairs per minute) faster in prokaryotes or eukaryotes?

Check Yourself

1. transforming substance, rough, smooth. **(DNA as the genetic material)**

2. a. Avery, MacLeod, and McCarty **(DNA as the genetic material)**

3. bacteriophage, bacteria **(DNA as the genetic material)**

4. sulfur, phosphorus **(DNA as the genetic material)**

5. c. Hershey and Chase **(DNA as the genetic material)**

6. a. stopped transformation; b. no effect; c. No effect. They found treatment of the S-strain transforming substance with DNase (an enzyme which breaks up DNA) prevented R-strain bacteria from turning into S-strain; RNase (cleaves RNA but not DNA) and protein-digesting enzymes had no affect and allowed transformation to occur. This pointed to DNA as the transforming substance. They also discovered that highly-purified S-strain DNA could cause transformation, another point of evidence. **(DNA replication in prokaryotes)**

7. b. Watson and Crick **(DNA as the genetic material)**

8. nucleotides, deoxyribose, phosphate, adenine, guanine, thymine, cytosine **(DNA as the genetic material)**

9. 32 percent adenine, 32 percent thymine, 18 percent guanine, 18 percent cytosine **(DNA structure)**

10. covalent, hydrogen **(DNA structure)**

11. hydrogen, A, T, C, G **(DNA structure)**

12. complementary **(DNA structure)**

13. a circle **(DNA replication in prokaryotes)**

14. The circular DNA is replicated. The parent strand and new strand are attached to the plasma membrane, and as the cell grows the circles of DNA are pulled apart until separated. **(DNA replication in prokaryotes)**

15. e. helicase **(DNA replication)**

16. a. RNA primer **(DNA replication)**

17. f. topoisoerase **(DNA replication)**

18. g. single-strand binding proteins **(DNA replication)**

19. b. DNA polymerase **(DNA replication)**

20. h. Okazaki fragments **(DNA replication)**

21. b. DNA polymerase **(DNA replication)**

22. d. template **(DNA replication)**

23. c. RNA polymerase **(DNA replication)**

24. b. DNA polymerase **(DNA replication)**

25. d. template **(DNA replication)**

26. In semiconservative replication, the new (daughter) double-stranded DNA molecules each have one strand from the parent DNA molecule and one newly-synthesized strand. If replication was conservative, one of the new double-stranded DNA molecules would be made entirely up of the parent DNA, while the other would consist of the two new strands. **(DNA replication)**

27. T A G G C T A G A T T T G C **(DNA replication)**

28. DNA, mRNA, rRNA, and tRNA molecules are made directly on the DNA template; protein is made indirectly from DNA through mRNA. **(DNA replication)**

29. a. True

    b. True

    c. False. Errors are rare, but do occur. DNA polymerase doesn't catch or correct all errors.

    d. False. DNA synthesis runs in a 5 to 3 direction on the new strand.

    e. False. The RNA primer is laid down by RNA polymerase, not DNA polymerase.

    f. False DNA polymerase, not RNA polymerase, removes the RNA primer. **(DNA replication)**

30. In prokaryotes, DNA replication occurs in the nucleoid region. In eukaryotes, DNA replication is carried out in the nucleus. **(DNA replication)**

31. b, d, c, a **(DNA replication)**

32. B-DNA, Z-DNA. DNA is known to exist in at least two forms in nature. The most common is B-DNA, described by the Watson and Crick model. The helix is in an alpha form and right-handed. Another form

is made up of a left-handed helix and this causes the helix to zig-zag in space. Hence the name Z-DNA **(DNA structure)**

33. False. Only about 70 percent of DNA is involved in protein synthesis. The function of the remaining 30 percent is not understood. **(DNA replication)**

34. transposons **(DNA as the genetic material)**

35. Adenine and guanine are purines, uracil and cytosine are pyrimidines. **(RNA: mRNA, tRNA, rRNA)**

36. DNA is made of nucleotides with the bases A, T, C, and G, and is double stranded. RNA is made up of the nucleotide bases A, U, C, G and is generally a single-stranded molecule. Additionally, the sugar is deoxyribose in DNA and ribose in RNA. **(RNA: mRNA, tRNA, rRNA)**

37. The 3 types of RNA are messenger RNA (mRNA), transfer RNA (tRNA), and ribosomal RNA (rRNA). **(RNA: mRNA, tRNA, rRNA)**

38. template, RNA polymerase **(RNA: mRNA, tRNA, rRNA)**

39. mRNA, mRNA **(RNA: mRNA, tRNA, rRNA)**

40. Two possible answers are correct: mRNA, transcription or rRNA, translation. **(RNA: mRNA, tRNA, rRNA)**

41. At least 20 tRNAs are needed to carry the 20 amino acids. **(RNA: mRNA, tRNA, rRNA)**

42. The tRNA is in the form of a cross, with the amino acid attached at one end and the anticodon on the other end. See your textbook for a detailed drawing. **(RNA: mRNA, tRNA, rRNA)**

43. RNA molecules are assembled on a DNA template through complementary base pairing. The tRNA molecule carrying the correct amino acid is aligned on the mRNA molecule by complementary base pairing between the tRNA anticodon and the mRNA codon. **(RNA: mRNA, tRNA, rRNA)**

44. (i) adds RNA primer to the DNA template to let DNA synthesis begin

    (ii) links nucleotides in forming RNA molecules. **(RNA: mRNA, tRNA, rRNA)**

45. Each group of three nucleotides is transcribed into one codon on the mRNA molecule. Codons are found on mRNA and it is the mRNA code that should be used in writing down the sequence of amino acids, using the codon table that is found in most biology textbooks. **(RNA: mRNA, tRNA, rRNA)**

46. There are 64 codons, since there are 64 ways of combining the four nucleotides in different orders. Two codons are for stopping the protein synthesis, 62 are for the 20 amino acids The codon for the amino acid methionine (AUG) is also the start codon. **(RNA: mRNA, tRNA, rRNA)**

47. a. 60 codons (180 nucleotides divided by 3 nucleotides per codon equals 60 codons). **(DNA structure)**

    b. Each codon on the mRNA molecule can code for one amino acid, and the entire molecule has codons for 60 amino acids. **(RNA: mRNA, tRNA, rRNA)**

48. Ribozymes—RNA with an enzymatic function—are known to exist in prokaryotes. They is the only exception to the rule that enzymes are proteins. **(RNA: mRNA, tRNA, rRNA)**

49. Ribosomes are smaller in prokaryotes. **(RNA: mRNA, tRNA, rRNA)**

50. 2, 3 **(RNA: mRNA, tRNA, rRNA)**

51. ribosome **(RNA: mRNA, tRNA, rRNA)**

52. DNA: T C T G A A T G A

    mRNA: A G A C U U A C U

    tRNA: U C U G A A U G A

    tRNA codons have the same sequence of nucleotides as the original DNA, except that in tRNA, U is substituted for T. **(RNA: mRNA, tRNA, rRNA)**

53. replication forks, replication bubble **(RNA: mRNA, tRNA, rRNA)**

54. Prokaryotes are faster than eukaryotes in replicating DNA. **(DNA replication)**

Grade Yourself

Circle the question numbers that you had incorrect. Then indicate the number of questions you missed. If you answered more than three questions incorrectly, you will then have to focus on that topic. (If a topic has less than three questions and you had at least one wrong, we suggest you study that topic also. Read your textbook, a review book, or ask your teacher for help.)

Subject: DNA Structure and Replication

Topic	Question Numbers	Number Incorrect
DNA as the genetic material	1, 2, 3, 4, 5, 7, 8, 34	
DNA replication in prokaryotes	6, 13, 14	
DNA replication	15, 16, 17, 18, 19, 20, 21, 22, 23, 24, 25, 26, 27, 28, 29, 30, 31, 33, 54	
DNA structure	9, 10, 11, 12, 32	
RNA: mRNA, tRNA, rRNA	35, 36, 37, 38, 39, 40, 41, 42, 43, 44, 45, 46, 47, 48, 49, 50, 51, 52, 53	

Protein Synthesis and the Control of Genes

Brief Yourself

The Genetic Code

DNA is written in one nucleotide code, and the message is transcribed into a complementary mRNA code. The genetic information is maintained in the DNA molecule and interpreted by the RNA into protein structure. The RNA code consists of 64 combinations of three of the four RNA nucleotides (A, U, C, G). Each set of three nucleotides is called a codon. Since there are only 20 amino acids involved in ordinary protein synthesis, the code turns out to be degenerate (redundant)—there is more than one codon for most amino acids. There are also several codons that indicate when to start and stop protein synthesis.

Transcription

DNA is used as a template on which the mRNA molecule is formed. During transcription, RNA polymerase moves down the DNA polymer, unwinding the double helix as it goes, and RNA nucleotides are laid down. Once the RNA strand as been assembled, the helix immediately rewinds. This mRNA, fresh from the DNA template, is primary mRNA. It must be processed to make mature RNA before it can be used.

Translation

Translation is the step in which the message is translated from the language of nucleotides into the language of amino acids—protein synthesis. A ribosome forms on a strand of mRNA. tRNA molecules bearing amino acids attach to the ribosome and a peptide chain is synthesized.

Gene mutations

Over time, changes (mutations) occur in the DNA code. Mistakes are made in replicating DNA, and the effect this has on the final protein depends on the nature of the change. If the change is made at the active site of an enzyme, the functioning of the protein may be drastically affected.

Genetic Control

The model for gene expression in prokaryotes is the operon, first described by Jacob and Monod as the *lac* operon. The operon is the unit of gene expression in prokaryotes. It consists of a regulator gene (which codes for a repressor molecule), promoter and operator regions, and structural genes which code for the enzymes, which in the case of the *lac* operon, break down lactose. In eukaryotes, rRNA is transcribed within the nucleolus, and the DNA has a special form. The DNA loops of eukaryotes contain nucleolar organizer regions, each consisting of multiple copies the gene responsible for rRNA transcription. Eukaryotic mRNA has long noncoding nucleotide sequences called introns (intervening sequences) which are spliced out in making mature mRNA. The retained portions are referred to as exons since they are expressed.

Test Yourself

Questions 1–6. Matching. Answers may be used once, more than once, or not at all.

Possible answers: Sanger, Watson and Crick, Beadle and Tatum, Meselson and Stahl, Pauling and Itano, Nirenberg and Matthaei.

1. One gene, one enzyme hypothesis.
2. One gene, one polypeptide hypothesis.
3. Decoded the mRNA codons.
4. Examined genes for a metabolic pathway in *Neurospora* bread mold.
5. Studied hemoglobin polypeptide chains in sickle-cell anemia.
6. Worked out the structure of the insulin molecule.
7. DNA contains the base _________ instead of ____________, which is found in RNA.
8. Using a table of mRNA codons found in most biology textbooks, find the codons for:

   (i) phenylalanine

   (ii) histidine

   (iii) alanine

   (iv) methionine
9. Write the peptide coded for by the following mRNA sequence:

   UUCUCUAGA
10. Write the peptide coded for by the following DNA sequence:

    CTACCAGAG
11. What DNA sequence coded for the following peptide?

    methionine-histidine-tyrosine
12. Identify the following as a sequence mRNA or DNA nucleotides, or as can't tell from the information given:

    (i) A A A G C G A

    (ii) A A A G C G T

    (iii) A A A G C G U
13. A student performs an experiment similar to the one done by Beadle and Tatum in the 1940s. The student is studying a metabolic pathway in a bacterium that can be summarized as follows:

    (q) (r) (s)

    compound F → compound G → compound H → compound I

    Where q, r, and s are enzymes catalyzing these reactions.

    The bacteria were treated by exposure to x-rays, and mutant strains of cells isolated. The results of an experiment with one strain is summarized below. A variety of culture media were tried with the following results.

    ("+" indicates growth, "-" indicates no growth).

	normal strain	mutant strain
plain agar	–	–
agar plus basic nutrients	+	–
agar, basic nutrients, compound F	+	–
agar, basic nutrients, compound G	+	–
agar, basic nutrients, compound H	+	+
agar, basic nutrients, compound I	+	+

Interpret the results and state which enzyme is defective in this particular mutation.

14. How was Sanger's determination of the primary structure of the protein insulin important for understanding the way DNA codes for proteins?

15. What is the central dogma of molecular biology?

    a. All enzymes are proteins.

    b. DNA is translated into mRNA and mRNA is transcribed into protein.

    c. DNA is transcribed into mRNA and then translated into protein.

    d. All biochemical reactions are governed by the same laws as ordinary, non-biochemical reactions.

16. True or false: mRNA is ready for translation as soon as it comes off the DNA template.

17. DNA codes for which of the following (pick all that are correct):

    a. Primary structure of a protein

    b. Secondary protein structure

    c. Tertiary protein structure

18. What components are needed to synthesize a protein?

19. How does a ribosome know when to stop making a polypeptide?

20. (i) What do the codons UAA, UAG, and UGA signify?

    (ii) What about codon AUG?

21. What is an initiation complex?

22. __________ enzymes (also known, more technically, as aminoacyl-tRNA synthetases) are responsible for binding the correct amino acid to a particular tRNA molecule. There are 6 different tRNA molecules that carry the amino acid leucine (each has a unique anticodon). Does this mean there are 6 different activating enzymes for leucine?

23. How many activating enzymes are there?

24. A ribosome has an ___ site, where incoming tRNA molecules first bind, and a ___ site where the polypeptide is attached.

25. The movement of the tRNA molecule from the ___ site to the ___ site is called __________. The movement of the ribosome along the mRNA strand is called __________.

26. Not all parts of a gene affect the final protein. Parts of the gene that are expressed are called ____ while the parts that are spliced out are called ______.

27. What are the three steps of protein synthesis?

28. Distinguish between a point and frameshift mutation.

29. Cells synthesize many proteins for export from the cell. True or false: Cells can only control protein function prior to the protein leaving the manufacturing cell.

30. What is an operon?

31. The concept of the operon is associated with the work of:

    a. Jacob and Monod

    b. Meselson and Stahl

    c. Beadle and Tatum

    d. Watson and Crick

32. Which of the following are parts of an operon: inducer, promotor, operator, structural gene (s), introns, regulator, repressor.

33. The *lac* operon is normally switched ____ (on or off) and is an example of an ______ operon. The *trp* operon, on the other hand is usually switched ___ (on or off) and is an example of an _______ operon.

34. In the *lac* operon, lactose is the __________. In the *trp* operon, tryptophan is the __________. Possible answers: repressor, corepressor, inducer, promoter.

35. When an *E. coli* bacterium is moved from a regular culture medium to one containing lactose as its only carbohydrate (carbon) source, is the *lac* operon turned on or off?

36. Operons are found in prokaryotes or eukaryotes?

37. How can gene amplification be used to meet immediate demands for a quantity of protein?

38. What is a Barr body?

39. The body of an adult human female is a mosaic in what sense?

40. What are polytene chromosomes, and where are they found? What are chromosome puffs?

41. What is the difference between primary and mature mRNA?

42. Distinguish between oncogenes and tumor suppressor genes.

43. Identify the following mutations as frameshift or point mutations.

    (i) ATE THE RAT becomes TET HER AT

    (ii) AUCCGA becomes AACCGA

44. What genetically-based blood disease results from the change in one amino acid in the hemoglobin molecule?

45. What is a mutagen?

46. __________ happens in the nucleus, whereas ______ happens in the cytoplasm. (Translation or transcription.)

47. Are the following statements true or false about cancer cells? If false, rewrite to make the statement true.

    (i) Cancer cells can only divide about 50 times.

    (ii) Cancer cells show contact inhibition.

    (iii) Cancer cells can spread or metastasize.

48. Will a change in the nucleotide sequence always effect a change in the protein?

49. How is gene expression controlled in eukaryotes?

Check Yourself

1. Beadle and Tatum **(The genetic code)**

2. Pauling and Itano. In the 1940s, Beadle and Tatum, experimenting with mutant strains of the red bread mold *Neurospora crassa*, established the hypothesis that a gene contains information for one enzyme (protein). This was later modified by Pauling and Itano to the currently accepted view, that a gene codes for one polypeptide. **(The genetic code)**

3. Nirenberg and Matthaei. Nirenberg and Matthaei worked out the code in which messenger RNA is written. **(The genetic code)**

4. Beadle and Tatum **(The genetic code)**

5. Pauling and Itano **(The genetic code)**

6. Sanger **(The genetic code)**

7. uracil, thymine **(The genetic code)**

8. (i) UUU, UUC

   (ii) CAU, CAC

   (iii) GCU, GCC, GCA, GCG.

   It is important to note that codons are complementary to the DNA nucleotide triplet on which the codon is formed, and that the codon table is for mRNA, not DNA. The codon for methionine (which is also the start codon for protein synthesis) is AUG. The DNA complement to this is TAC. **(The genetic code)**

9. phenylalanine-serine-arginine **(The genetic code)**

10. aspartate-glycine-leucine (mRNA is GAU GGU CUC) **(The genetic code)**

11. methionine is TAC, histidine is either GTA or GTG, and tyrosine is either ATA or ATG **(The genetic code)**

12. (i) Can't tell—A, G, and C are found in both DNA and mRNA.

    (ii) DNA is indicated by the presence of T.

    (iii) mRNA is indicated by the presence of U. **(The genetic code)**

13. The mutant strain can't synthesize the product that is made by enzyme r, and perhaps lacks a functional enzyme q as well. That means compound H and compound I are not being made and must be supplied from an outside source. When either of these compounds is supplied in the medium the mutant strain can grow normally. **(The genetic code)**

14. The determination of the primary structure of insulin showed that proteins had a definite composition in terms of specific amino acids in a specific order. This suggested that genes would likewise have a highly ordered structure. **(The genetic code)**

15. c. DNA contains the information for the primary structure of protein, which is transcribed into mRNA and translated into protein. Information flows from DNA to mRNA to protein, and not in the other direction. **(The genetic code)**

16. False. Freshly-synthesized mRNA is primary mRNA—it must be processed to make mature mRNA. **(Transcription)**

17. a. DNA codes only for the primary structure of proteins; the other levels of protein organization are a consequence of that primary sequence. **(Transcription)**

18. amino acids, tRNA, mRNA transcript, ribosome components, enzymes, source of energy. DNA is required for the initial nucleotide code for the protein. **(Translation)**

19. Stop codons, for which there are no corresponding tRNA molecules, terminate protein synthesis. **(Translation)**

20. (i) These are the stop codons.

    (ii) AUG is the start codon and also the codon for methionine. **(Translation)**

21. Ribosome, tRNA molecule, mRNA molecule. **(Translation)**

22. activating. There is only one activating enzyme for each amino acid, even if there are multiple tRNA molecules (each with a different anticodon) for that amino acid. **(Translation)**

23. There are 20 enzymes, one for each amino acid. There are not 64 enzymes (see answer directly above). **(Translation)**

24. A, P **(Translation)**

25. A, P, translocation, elongation. Explanation: At a given time, two tRNA molecules can be found on the ribosome; one occupies the A site and holds an amino acid, the other is attached to the P site, and holds on to the growing peptide chain. During translocation, the tRNAs shift, the one at the P site leaving the ribosome and the other shifting from the A site to the P site. This latter tRNA molecule acquires the peptide chain, adding its own amino acid to it. Meanwhile, the entire ribosome has also moved relative to the mRNA molecule it is reading, and a new tRNA molecule moves into the vacant A site.

    In working out this process, it is helpful to envision the ribosome as having two chairs, the A and P sites, each occupied by a child (tRNA molecule). The child in the A chair has a single balloon (amino acid) and the child in the P chair has a string of balloons. The children shift to one side, one child leaving the ribosome but not before passing the string of balloons to the other child, who adds his or her balloon to it. Another child moves into the vacant A chair, and the process goes on. The tRNA molecules match up a nucleotide triplet (the anticodon) with each codon on the mRNA, ensuring the correct amino acid is added. This is another example of complementary base pairing, only in this case it is mRNA and tRNA nucleotides that are matching up. If the codon is ACG, the anticodon is UGC. Note that the anticodon is a copy of the DNA triplet (in this case TGC) except that U is substituted for T in DNA. **(Translation)**

26. exons, introns. In converting the initial mRNA strand to mature mRNA. Remember this: *ex*ons are *ex*pressed. The introns are removed and the exons are spliced together. **(Transcription)**

27. Initiation (of synthesis), peptide chain elongation, peptide chain termination **(Translation)**

28. In a *point mutation,* one nucleotide is changed for another. Due to the redundancy of the genetic code, this may make no difference. For instance, if (speaking in mRNA codon terms) AUU changes to AUC, it doesn't matter. Both codons code for isoleucine. But if AUU changes to AUG, a methionine will be substituted for the isoleucine. What difference such a substitution makes depends on the differences between the amino acids and where the amino acids are found in the protein. Another type of mutation, called *frameshift,* usually result in a nonfunctional protein. In this type of mutation, a nucleotide is deleted

or a new nucleotide inserted. This changes the groupings of three nucleotides which are read as codons. Everything that occurs after the mutation is altered. **(Gene mutations)**

29. False. Proteins such as enzymes may, for instance, be secreted from a cell in an inactive form (termed zymogens). The digestive enzyme pepsin is secreted as pepsinogen, and in the presence of hydrochloric acid (as in the stomach) loses a short, terminal peptide chain. This activates the enzyme. Of course, it is other cells, not the manufacuring cell which exerts this post-secretion control. **(Gene mutations)**

30. The operon is a unit of gene function. **(Genetic control)**

31. a. Jacob and Monod **(Genetic control)**

32. promoter, operator, structural gene (s), regulator **(Genetic control)**

33. off, inducible, on, repressible. Normally, the repressor molecule binds to the DNA at the operator site, blocking RNA polymerase from gaining access to the DNA. When lactose is present in the culture medium, it binds to the repressor, inactivating it. The adjacent promoter site is now accessible to RNA polymerase and transcription of the structural genes can start. The *lac* operon is normally off and is termed inducible, that is, it can be switched on when an inducer (lactose) binds to the repressor. Another type of operon, such as the *trp* (tryptophan) operon is by contrast normally switched on. When tryptophan binds to a corepressor molecule, this in turn can bind to the operator, blocking the promoter region and preventing transcription of the structural genes. **(Genetic control)**

34. inducer, repressor. In both instances the molecule in question (lactose or tryptophan) affect the expression of the operon. In the *lac* operon, lactose binds with the repressor, which then binds to the operator region, and allows a normally switched-off gene to be switched on. In the *trp* operon, tryptophan binds to the repressor and this in turn binds to the operator region and prevent the expression of a normally switched-on operon. **(Genetic control)**

35. The *lac* operon is turned on when lactose is present and must be metabolized. **(Genetic control)**

36. prokaryotes **(Genetic control)**

37. Gene amplification is a way of gearing up production to meet a demand for a protein. The relevant section of DNA is replicated again and again, so multiple copies of the genes are available for transcription. **(Genetic control)**

38. In female mammals, the Barr body is a condensed, inactive X chromosome. All cells of female mammals, except for some found in the ovaries, exhibit Barr bodies. **(Genetic control)**

39. A human female has one X chromosome from her mother and one from her father. Which one of the pair of X-chromosomes will be inactive in a particular cell is a matter of chance. Therefore, one cell may express alleles on the paternally-derived X-chromosome, while a neighboring cell expresses alleles on the maternally-derived X-chromosome. It can be generalized that in female mammals, only one of the two X-chromosomes is expressed in the average body cell; which one is a matter of chance. This leads to there being differences between cells depending on which alleles are expressed in that cell. **(Genetic control)**

40. Polytene chromosomes are chromosomes that have been replicated again and again without division. They are found in the salivary gland cells of fruit flies and related insects. Chromosome puffs are regions of chromosomes that bulge, and are being actively transcribed into mRNA. **(Genetic control)**

41. Primary mRNA has just come off of the DNA template strand. After it is given a cap and a poly A (polyadenine) tail and the introns are spliced out by spliceosomes, the mRNA is referred to as mature. **(Transcription)**

42. Oncogenes contain information for the production of proteins that cause cells to divide in an uncontrolled (cancerous) fashion. Tumor suppressor genes makes proteins that have the opposite effect. **(Cancer)**

43. (i) the addition (or deletion) of a nucleotide identifies this as a frameshift mutation. After the point of deletion or insertion, the reading frame is displaced and new nucleotide triplets are recognized (incorrectly) as codons)

    (ii) A substitution of one nucleotide for another is a point mutation. **(Gene mutations)**

44. Sickle-cell anemia. Pauling and Itano examined hemoglobin from people homozygous for the recessive allele that causes sickle-cell anemia using electrophoresis. They compared their hemoglobin to that of people homozygous for the normal allele and also to people who were heterozygous (with one sickle-cell and one normal allele). The results showed that sickle-cell anemia is due to a modification in the protein structure of hemoglobin. **(Gene mutations)**

45. A mutagen is a substance (or source of energy) which changes the nucleotide composition of DNA in cells. **(Gene mutations)**

46. Transcription, translation **(The genetic code)**

47. (i) False. Normal cells can only divide about 50 times. (Sometimes called the Hayflick limit.) When cells are dividing uncontrollably, and their growth is no longer inhibited by contact with other cells, or limited (as in normal cells) to a lifespan of about 50 divisions, the cells have become cancer cells. These abnormal cells grow rapidly, move from one part of the body to another, and attempt to monopolize resources. Some cancerous cells have been shown to differ from normal cells with respect to one gene, an oncogene.

    (ii) False. Normal cells show contact inhibition.

    (iii) True. **(Cancer)**

48. No. A change in a nucleotide (or several nucleotides) may have no effect if the new codon codes for the same amino acid (e.g., if UUU mutates to UUC, since both are codons for phenylalanine) or even for a very similar amino acid (e.g., if GGG becomes GCG, glycine will be replaced by a the very similar alanine. **(Gene mutations)**

49. Control of gene expression in eukaryotes is more complex and not as well understood as gene expression in prokaryotes. It is known that some forms of eukaryotic chromatin are transcribed, while other forms are not. Additionally, DNA in eukaryotic cells is found in several forms: as a single copy (one or a few copies of the sequence exist); or intermediate repeat sequences (as much as 40 percent of DNA is formed of such repeating sequences); and simple sequence DNA, which is short but has highly repetitive sections. The significance of these various forms of DNA is still disputed. **(Genetic control)**

Grade Yourself

Circle the question numbers that you had incorrect. Then indicate the number of questions you missed. If you answered more than three questions incorrectly, you will then have to focus on that topic. (If a topic has less than three questions and you had at least one wrong, we suggest you study that topic also. Read your textbook, a review book, or ask your teacher for help.)

Subject: Protein Synthesis and the Control of Genes

Topic	Question Numbers	Number Incorrect
The genetic code	1, 2, 3, 4, 5, 6, 7, 8, 9, 10, 11, 12, 13, 14, 15, 46	
Transcription	16, 17, 26, 41	
Translation	18, 19, 20, 21, 22, 23, 24, 25, 27	
Gene mutations	28, 29, 43, 44, 45, 48	
Genetic control	30, 31, 32, 33, 34, 35, 36, 37, 38, 39, 40, 49	
Cancer	42, 47	

Recombinant DNA, Biotechnology, and Genetic Engineering

Brief Yourself

Gene Cloning and PCR

DNA cloning and polymerase chain reaction (PCR) methods are used to make many copies of a particular DNA sequence. In cloning, a vector such as a virus is used to carry the DNA segment into a bacterial cell, where the vector replicates, making more of the DNA sequence at the same time. In the PCR technique, DNA polymerase is used to make copies of a sequence once primers (short complementary nucleotide sequences) have been added and have found their place on each end of the strand segment to be copied. The DNA polymerase fills in the nucleotides in the gap between the primers. This process is repeated a number of times to give large numbers of copies. This is aptly termed amplification.

Biotechnology

Biotechnology is a new field in which the genomes of plants, animals (even humans) are manipulated. For instance, the human gene for insulin has been spliced into a bacterial genome, and the bacteria now produce human insulin in addition to native bacterial proteins.

Recombination

Recombination is the process by which DNA from two (or more) sources is combined. The result is rDNA (recombinant DNA). A vector, such as a virus or bacterial plasmid (small circular piece of DNA separate from the main genome), is often used to carry DNA from one source into another genome. Recombination is also a natural phenomenon among bacteria. Bacteria can transfer genetic elements (such as for antibiotic resistance) among themselves by the processes of transduction, transformation, and conjugation.

Test Yourself

1. What is recombinant DNA?

2. What are plasmids?
   a. The circular prokaryotic "chromosome"
   b. Eukaryotic genes
   c. A small ring of DNA found in bacteria
   d. A virus used as a vector

3. How did restriction enzymes get their name?

4. Are genetic experiments a uniquely human activity?

5. What are the three basic steps in moving a gene from one organism to another?

6. What is complementary DNA (cDNA) and how is it made?

7. True or false: a DNA probe is single strand of DNA that binds to a nucleotide sequence of interest, such as a specific gene.

Questions 8-11 . Identify the following as transduction, transformation, or conjugation.

8. Bacterial cell takes up DNA from medium, similar to Griffth's early experiment.

9. An F plasmid opens up, and a single strand moves into a recipient cell.

10. A bacterial virus introduces new bacterial genes into a cell.

11. Hfr, an integrated plamid, brings DNA from the host genome into the recipient cell.

12. What are four products of biotechnology?

13. PCR is an abbreviation for ________________.

14. What happens to double stranded DNA if it is gently heated?

15. How are specific DNA sequences located?

16. What are some uses of PCR techniques?

Questions 17–26. Matching. Answers are used once or not at all.

Possible answers: plasmid, bacteriophage, transgenic, Ti plasmid, retrovirus, gene therapy, DNA marker, cDNA, rDNA, sDNA, Human Genome Project, DNA polymerase, DNA ligase, restriction enzymes.

17. Providing healthy or normal genes to the body to replace defective ones.

18. An RNA virus that carries out reverse transcription.

19. A repeating pattern of DNA sequences that is highly individual—the basis of DNA fingerprinting.

20. DNA combined from 2 or more sources.

21. Enzymes which restrict the growth of viruses.

22. An enzyme that closes up gaps in a DNA molecule.

23. A ring of bacterial DNA.

24. DNA that does not contain introns.

25. Mapping and sequence the human chromosomes.

26. Vector for moving genes into plant cells.

27. What are transgenic organisms?

28. What are some possible applications of transgenic engineering in plants?

29. Antibiotic resistance, a growing world-wide problem, can be spread from one bacterial strain to (list all that apply): (i) other bacteria of the same strain; (ii) other bacteria of the same species; (iii) bacteria of closely related species; (iv) virtually any bacteria.

30. Why should we be concerned about the overuse of antibiotics in treating minor illnesses and as additives to animal feed?

31. Where in the cell are the genes for bacterial resistance to antibiotics found—the plasmid or the main, circular bacterial DNA ("bacterial chromosome")?

32. What enzymes split DNA strands into fragments with "sticky ends"?

Questions 33–36.

A student is experimenting with two strains of the bacterium *Escherichia coli* (*E. coli*). One strain, identified as J-53R, is susceptible to the antibiotic chloramphenicol, but is resistant to the antibiotic rifampicin. Another strain, HT-99 shows the opposite pattern; it is resistant to chloramphenicol but susceptible to rifampicin. The resistance gene in the latter is located on a plasmid in the cell. It is not part of the main bacterial genome.

The student inoculated the following media with bacteria. The mixed bacteria consisted of both strains mixed together 24-hours prior to being put on the media. ("+" indicates, growth; "-" indicates no growth.)

Bacterial strain	Chloramphenicol medium	Rifampicin medium	Medium with both antibiotics
HT-99	+	–	–
J-53R	–	+	–
mixed strains	+	+	+

33. Which strains were resistant to which antibiotic according to the results above?

34. Why was growth obtained on the medium with both antibiotics only after the strains were mixed?

35. By what process were genes passed from cells of one strain to cells of another?

36. Does the experimental evidence support the conclusion that antibiotic resistance was transferred?

37. Where in a DNA molecule does a restriction enzyme break bonds?

38. What hereditary disease can now be diagnosed with good reliability, if not 100 percent certainty, using DNA from individuals?

39. What is the Human Genome Project?

Check Yourself

1. Recombinant DNA (rDNA) is DNA from two different sources that is put together in a test tube ("*in vitro*"). **(Recombination)**

2. c. A small ring of DNA found in bacteria **(Gene cloning and PCR)**

3. Restriction enzymes are so-called because they restrict the growth of viruses. They are used to cut the DNA molecule open. Such an enzyme cuts at a specific place. "Sticky-ended" DNA results, and foreign DNA can be inserted. DNA ligase closes the gap, sealing in the foreign DNA. **(Recombination)**

4. No. Genetic experiments have been and are being carried out in the natural world during the evolution of life. What is unique about biotechnology is that it is directed by humans. **(Biotechnology)**

5. The three steps are: (1) isolate the DNA sequence of interest, (2) cut it out and (3) finally splice it into the desired target genome. Eukaryotic genes need processing before they are inserted into bacteria. Introns must be removed form the DNA, for example. This can be done by synthesizing the DNA segment without the introns, or making the new DNA on a strand of complementary mature mRNA, from which the introns have been removed. **(Recombination)**

6. Complementary DNA (cDNA) is made on a mature mRNA template. cDNA does not contain introns, since intron sequences have been spliced out of mature mRNA. **(Recombination)**

7. True **(Biotechnology)**

8. Transformation **(Recombination)**

9. Conjugation **(Recombination)**

10. Transduction **(Recombination)**

11. Conjugation **(Recombination)**

12. The human hormones somatostatin, insulin, and growth hormone, and the enzymes renin (used in cheese making) and cellulase. **(Biotechnology)**

13. Polymerase chain reaction **(Gene cloning and PCR)**

14. The strands separate as the hydrogen bonds between complementary base pairs are broken. On cooling, the hydrogen bonds spontaneously reform. **(Gene cloning and PCR)**

15. DNA is heated, separating the strands. Fragments of DNA made up of sequences complementary to the regions we are interested in are added, and these bind to complementary sections as the DNA cools. The amount of base pairing is reflected in the extent and rate at which the double-stranded DNA is formed. **(Biotechnology)**

16. PCR is used to make multiple copies of segments of DNA. Using only a small amount of original DNA, many copies (millions) can be made in a days' time. PCR techniques are widely employed in society, for example in crime detection to help identify rapists, in the study of genetic diseases and in investigating the evolutionary relationships of various organisms by comparing their DNA. It is even possible to come

up with information about roughly how many years ago two species (such as humans and chimpanzees) shared a common ancestor. **(Gene cloning and PCR)**

17. gene therapy **(Biotechnology)**

18. retrovirus **(Biotechnology)**

19. DNA marker **(Biotechnology)**

20. rDNA **(Recombination)**

21. restriction enzymes **(Recombination)**

22. DNA ligase **(Recombination)**

23. plasmid **(Recombination)**

24. cDNA **(Recombination)**

25. Human Genome Project **(Biotechnology)**

26. Ti plasmid **(Biotechnology)**

27. Transgenic organisms are free-living, natural organisms that have a foreign piece of DNA (a gene) inserted into their genome. Eventually improved breeds of milk and meat animals and varieties of crops, all produced by genetic engineering, will replace those developed over years by farmers using traditional methods of artificial selection in animal and plant husbandry. **(Gene cloning and PCR)**

28. Plants might be developed that are more resistant to insects, disease,s and herbicides, or that give an increased crop yield. **(Recombination)**

29. i, ii, iii, and iv are all true. **(Recombination)**

30. Such practices result in genes for antibiotic resistance becoming common in bacteria, which adversely affects public health. **(Recombination)**

31. Genes for antibiotic resistance can be found in both places, but those borne on plasmids are most readily transferred to other cells and thus a greater threat. **(Recombination)**

32. restriction enzymes **(Recombination)**

33. The HT-99 strain is resistant to chloramphenicol, and the J-53R strain is resistant to rifampicin. **(Recombination)**

34. After the strains were mixed the plasmid from the HT-99 strain migrated to the J-53R strain, producing a strain resistant to both antibiotics. **(Recombination)**

35. conjugation **(Recombination)**

36. Yes, the results are consistent with that hypothesis. **(Recombination)**

37. The restriction enzyme breaks the covalent bond link between the sugar and phosphate groups. **(Recombination)**

38. Huntington's disease (Huntington's chorea) **(Biotechnology)**

39. The Human Genome Project is an international, large-scale scientific venture that plans to map the nucleotide sequence on the human chromosomes in the next 10 years (or sooner). **(Biotechnology)**

Grade Yourself

Circle the question numbers that you had incorrect. Then indicate the number of questions you missed. If you answered more than three questions incorrectly, you will then have to focus on that topic. (If a topic has less than three questions and you had at least one wrong, we suggest you study that topic also. Read your textbook, a review book, or ask your teacher for help.)

Subject: Recombinant DNA, Biotechnology, and Genetic Engineering

Topic	Question Numbers	Number Incorrect
Recombination	1, 3, 5, 6, 8, 9, 10, 11, 20, 21, 22, 23, 24, 28, 29, 30, 31, 32, 33, 34, 35, 36, 37	
Gene cloning and PCR	2, 13, 14, 16, 27	
Biotechnology	4, 7, 12, 15, 17, 18, 19, 25, 26, 38, 39	

Evolution

12

Brief Yourself

Darwin and Natural Selection

Evolution is the central theme of modern biology. No other idea touches on all aspects of the study of life. While Darwin did not invent or discover the idea of evolution, he and Wallace were the first to propose a workable mechanism for how evolution happens. Building on their knowledge of the variation found among individual organisms in nature, they saw that the struggle for survival would tend to favor the best adapted individuals, and that those individuals would have more descendants. Over generations, repeated selection for particular characteristics would increase the number of individuals with those characteristics in a population relative to the rest of the population.

Evolution begins with changes in the gene pool of populations, not in individual organisms. Genetic changes occur through the process of mutation, and changes in allele frequencies occur through natural selection and random events.

Genetics and Evolution

Genetics underlies evolution, although selection works on the phenotype and not directly on the genotype. Forces of natural selection include isolation of populations with accompanying sampling errors, mutations which change genes, and competition between individuals for limited resources. According to the mathematics of the Hardy-Weinberg equation, the allelic (gene) frequencies with a population will remain constant in the absence of outside or disruptive forces (such as mutation or genetic drift, gene migration, etc.) as long as mating is random. Hence, when a change in allelic frequency is seen, it must be due to some selective force at work.

Evolution operates in a specific environment. One of the most satisfying evidences of evolution is to examine organisms from similar environments that are isolated by geography from direct contact, and find that similar selective forces—due to like environments—produce similar adaptations.

Microevolution and Macroevolution

Evolutionary changes can be viewed on a larger scale (often from the fossil record) as macroevolution, or with a concentration on the forces producing change (microevolution). Macroevolutionary changes occur above the species level, and must be viewed within the framework of geologic time. Microevolution—evolutionary changes within populations—are the results of molecular level shifts and lead to variations within species.

Test Yourself

1. __________ was a predecessor of Darwin's and developed the theory of acquired characteristics.

2. Darwin's book, *On the Origin of Species*, was first published in __________.

3. What was the generally accepted explanation for organisms and their characteristics prior to Darwin and Wallace?

4. Lamarck explained the giraffe's long neck as a result of what?

5. What is a fossil? How are fossils evidence of evolution?

6. What are the main points of Darwin's theory of natural selection?

7. Darwin made a five-year voyage on board the ship the *H.M.S.* __________.

8. How would Darwin explain the giraffe's long neck?

9. Comparative anatomy is the study of what? How can comparative anatomy provide evidence for evolution?

10. What is another, modern way of testing the relationship between two species?

11. Give an example of an analogous structure and a homologous structure.

12. Over decades, the peppered moth changed from a light-colored to a dark-colored moth near industrial Birmingham, in England. What is this an example of? What is the agent of selection?

13. What happens to most species over time?

14. Darwin lacked a basic tool for understanding evolution, a knowledge of __________.

15. The fundamental unit of natural selection is the __________, not the __________.

16. All the genes present in a population at one time constitute the __________.

    a. genome

    b. gene pool

    c. gene reservoir

    d. genetic plan

17. Darwin was greatly influenced in his thinking by reading a book ("essay") by __________.

    a. Haldane

    b. Wallace

    c. Lamarck

    d. Malthus

18. Is survival random according to Darwin? If not, what determines it?

19. What is fitness, and how is it measured?

20. What are the main tenets of the Hardy-Weinberg equation?

21. What is genetic drift?

22. How does isolation of species affect evolution? In what ways can species be isolated?

23. What is the "ultimate source of novelty" in evolution?

24. What is microevolution, and how does it differ from macroevolution?

25. What are some sources of genetic variation in individuals?

26. Of what value is heterozygosity in evolution? Use the example of sickle cell anemia in your answer.

27. Directional selection refers to what?

28. What is stabilizing selection?

29. How can reproductive isolation come about?

30. Major differences in the phenotype between the sexes are examples of _________ ______. (two words)

31. What is the effect of stabilizing selection on variation in a population?

32. Does selection act directly on the genotype or the phenotype?

33. When two species are formed from a single ancestral species, this is called ____________ evolution.

34. Two species of trees may not interbreed because they release pollen at different times of the year. This is an example of:

    a. temporal isolation

    b. geographic isolation

    c. gametic isolation

35. Animals evolve elaborate courtship rituals under what conditions?

36. What is prezygotic isolation?

37. Give an example of a post-zygotic isolating factor.

38. Allopatric speciation occurs when populations are:

    a. socially isolated

    b. geographically isolated

    c. isolated by mutation from the parent population

    d. breeding at a different time of the year from the parent population.

39. Briefly explain adaptive radiation and give a simple example.

40. What is the theory of punctuated equilibrium?

41. Humans are thought to have evolved _____ millions years ago in __________.

42. Humans diverged from a common ancestor with our closest primate relative, the _____________ between ______ and ________ million years ago.

43. Humans belong to what mammalian order?

44. The earliest hominid is thought to have been _____________.

45. In a small town insecticide has been widely sprayed for a number of years to control mosquitoes. The results the first few summers were outstanding: almost no mosquitoes. But each summer thereafter, the number of insects increased until eventually they were as numerous as before the spraying began. What happened? Explain with reference to natural selection.

46. There are many remarkable similarities between eutherian (placcntal) mammals in South America and marsupial (non-placental) mammals found in Australia. How can these similarities be accounted for in evolutionary terms?

47. Does the fact that both birds and bats have wings suggest they have a winged ancestor in common?

48. Some fish that live in caves are blind. How can blindness—ostensibly a defect—be explained as an adaptation?

Check Yourself

1. Lamarck **(Darwin and natural selection)**

2. 1859 **(Darwin and natural selection)**

3. The accepted explanation was special creation—that God created each species by an individual act of creation. **(Darwin and natural selection)**

4. Lamarck would have argued that according to his theory of acquired characteristics, as a giraffes reached higher and higher into trees to get to leaves during times of intense competition for food, they would have stretched their necks. This stretching would have affected the neck lengths of their offspring, making them longer bit by bit. Repeated over many years, the effect would produce animals with very long necks. **(Darwin and natural selection)**

5. Fossils are found in sedimentary rocks and are the mineralized remains of earlier life. Examination of fossils has revealed changes in various groups of animals over geological time. **(Darwin and natural selection)**

6. The main points of Darwin's theory are:

   (i) organisms vary

   (ii) some of this variation is inherited

   (iii) some variations cause an organism to be able to better cope with environmental demands

   (iv) organisms produce far more offspring than can survive

   (v) in the struggle for existence which ensues, the organisms with variations which make them better suited to survive and reproduce will have more viable offspring (differential reproductive success).

   Cast in modern terms, the theory of evolution can be stated this way:

   (i) the process of reproduction is stable

   (ii) there are chance variations among individual organisms, and some are inheritable

   (iii) the number of individuals that survive is small compared to the number produced

   (iv) survival is determined by interactions between chance variations and the environment. **(Natural Selection)**

7. *Beagle* **(Darwin and natural selection)**

8. Darwin would have said that the giraffe's with even slightly longer necks (a result of natural variation in neck lengths) could reach higher into the trees for leaves. They would tend to have more offspring in each generation since they have a better chance to survive and reproduce. If the process repeats itself over many generations, even a slight advantage in having a longer neck will select in favor of a giraffe with a long neck. Like Lamarckian evolution, the process moves slowly, but the driving force is different in each theory. **(Darwin and natural selection)**

9. Comparative anatomy is the study of similarities and differences in the structure of animals. By demonstrating similarities of structure between two different species of animals, a common ancestor with that structure can be postulated. Lacking a close enough relationship, similar selective pressures may be envisioned as producing similar results in both instances. For example, Australia has marsupial animals

which correspond in structure and life habits to eutherian (placental) mammals in similar environments on other continents; and although these animals each evolved apart from one another due to continental drift, similar environmental pressures produced similar animals. **(Darwin and natural selection)**

10. Comparing DNA sequences is another way of determining the closeness of relationship between two species. **(Darwin and natural selection)**

11. Analogous structures have similar functions; e.g., the wings of bats, flying squirrels, and insects. All are used for flying or gliding, but were evolved separately. Homologous structures have a common origin in development and evolution—the fins of porpoises and legs of a dog are both derived from limbs in the basic vertebrate body plan. Some structures may be both analogous and homologous (wings of birds and bats are both modifications of the front limbs.) **(Darwin and natural selection)**

12. As the tree trunks on which the moths rest became darker from pollution, it became more and more of an advantage for others to have darker coloration. Within the range of coloration found in the natural population, selection favored those that were darker. The agent of selection in this case were the birds who more readily picked out and consumed the lighter colored moths, leaving more of the darker moths to reproduce. **(Darwin and natural selection)**

13. Most species have, and presumably will continue, to become extinct over long periods of time. **(Darwin and natural selection)**

14. inheritance or genetics **(Darwin and natural selection)**

15. population, individual **(Darwin and natural selection)**

16. b. gene pool **(Genetics and evolution)**

17. d. Darwin opens his *Origin* book by referring to reading Malthus' book-length essay on population, and further acknowledges his debt in his *Autobiography*. **(Darwin and natural selection)**

18. Survival is not generally a random outcome, but a result of competition between the better and less well adapted individuals in a population. Superior (better-adapted) organisms are more likely to survive and reproduce. Natural selection determines who will survive. **(Darwin and natural selection)**

19. Fitness has been described in many ways, but one of the most widely-accepted is to state it in terms of the number of offspring and descendants an individual contributes to the next and subsequent generations. More fit organisms will have relatively more of their genes represented in future generations. **(Darwin and natural selection)**

20. The Hardy-Weinberg equation is a mathematical way of saying that the frequency with which a gene (an allele) is found in a population will remain constant over time if the mating is random and in the absence of outside forces. That is, allelic frequencies will remain stable unless acted upon by selective pressures such as gene flow, mutation, genetic drift, nonrandom mating (as in sexual selection). **(Darwin and natural selection)**

21. Genetic drift is a sampling error. When a small group of organisms from a population become isolated, they found another population based only on the genes (alleles) they carry with them, which is less than the total number of alleles found in the parent population. These organisms are only a partial genetic sample, not representative of the entire populations gene pool. **(Darwin and natural selection)**

22. Isolation is one way in which populations within a species can evolve separately from the parent population and eventually evolve into a new species. Ways of becoming isolated include mutations which change timing of reproduction (temporal isolation) or isolation enforced by geography—separated on an island or by mountains. **(Darwin and natural selection)**

23. Crossing over and independent assortment of chromosomes (sex) increase the genetic variability between individuals, and provide the great bulk of variation in evolution; but in an ultimate sense, the only true source of new nucleotide sequences—or genes—is mutation. **(Genetics and evolution)**

24. Microevolution are the small changes in gene frequencies caused by mutation, genetic drift, natural selection, and similar evolutionary agents. Macroevolution refers to large scale changes in groups over periods of geological time. **(Darwin and natural selection)**

25. Crossing over, independent assortment of chromosomes, mutation **(Genetics and evolution)**

26. Recessive alleles are preserved by heterozygotes and also protected from exposure to natural selection. In sickle cell anemia, the sickle cell allele is advantageous in the heterozygote (a carrier), lethal in the homozygous recessive. **(Genetics and evolution)**

27. In response to a new or modified environment, any evolution that shows a trend is considered to be directional.There is a direction to the change in allelic or gene frequencies. **(Darwin and natural selection)**

28. Stabilizing selection tends to keep alleleic frequencies constant, as in a steady, relatively constant environment. This slows down evolutionary change. **(Microevolution and macroevolution)**

29. Reproductive isolation can come about by chance separation of subgroups from the parent population. A mutation can change breeding habits and effectively isolate the parent and subgroups, or the subgroup can become physically isolated, as would happen if migrating birds were blown off course to an isolated island. **(Darwin and natural selection)**

30. sexual dimorphism **(Darwin and natural selection)**

31. convergence **(Microevolution and macroevolution)**

32. Phenotype, but ultimately it will favor certain genotypes and change their frequency in a population. **(Darwin and natural selection)**

33. divergent **(Microevolution and macroevolution)**

34. a. temporal isolation **(Microevolution and macroevolution)**

35. Intense competition for mates will tend to produce courtship rituals of increasing complexity. **(Darwin and natural selection)**

36. This refers to characteristics that prevent fertilization from happening (hence *pre*-zygotic), such as incompatible courtship rituals. **(Microevolution and macroevolution)**

37. These factors include any that occur after fertilization, such as offspring being sterile as when a horse is crossed with a donkey giving a sterile mule. **(Microevolution and macroevolution)**

38. b. geographically isolated **(Microevolution and macroevolution)**

39. Adaptive radiation occurs when different lineages branch away from an ancestral group—mammals evolved into many different orders form a common ancestral group. **(Microevolution and macroevolution)**

40. The Gould-Elderidge theory of punctuated equilibrium suggests evolution often occurs in large steps—periods of intense evolutionary change occurring between periods that are relatively quiet and unchanging. This is often contrasted to what is perceived to be Darwin's approach, that evolution occurs in many small steps over long periods, and proceeds at a fairly uniform rate over time. **(Darwin and natural selection)**

41. five, Africa **(Human evolution)**

42. chimpanzee, 5, 8 **(Human evolution)**

43. Primates **(Human evolution)**

44. *Homo habilis* **(Human evolution)**

45. Initially, most of the mosquitoes were susceptible to the insecticide and died. A small group of insects, however, were resistant to the chemical's effect. This resistance was inheritable in at least some instances. These mosquitoes not only survived the spraying, but no longer had to compete with a large number of other mosquitoes (since those were killed by the spraying), which gave them a considerable advantage in terms of producing offspring. At first their numbers were small, but they increased each year until the mosquito population consisted of resistant individuals. In a sense, the insecticide had selected for exactly those individuals most able to resist its effects. **(Darwin and natural selection)**

46. While these two groups of mammals have been isolated for millions of years, and evolved independently of each other, they lived and evolved in similar environments. Similar environments place similar demands on organisms, and the adaptations which are favored in response will also be similar. **(Darwin and natural selection)**

47. No. As was the case in question 46, the fact that each group has evolved wings and flight behavior simply means they were subjected to similar selective pressures during their evolutionary history. This reasoning must be applied with care, however. Insects also have evolved wings, but these probably originated as a mean of thermoregulation and only later as a mechanism for flight. **(Darwin and natural selection)**

48. Blind cave fish have evolved under conditions of complete darkness. In such circumstances there is no advantage to having vision. Indeed, there is a disadvantage—energy is required during development to make eyes. Blind individuals were at no disadvantage compared to fish with eyes, and may have gained a slight advantage by not putting energy into the production of eyes. That is, eyes were not selected for, and may have even been selected against. **(Darwin and natural selection)**

Grade Yourself

Circle the question numbers that you had incorrect. Then indicate the number of questions you missed. If you answered more than three questions incorrectly, you will then have to focus on that topic. (If a topic has less than three questions and you had at least one wrong, we suggest you study that topic also. Read your textbook, a review book, or ask your teacher for help.)

Subject: Evolution

Topic	Question Numbers	Number Incorrect
Darwin and natural selection	1, 2, 3, 4, 5, 6, 7, 8, 9, 10, 11, 12, 13, 14, 15, 17, 18, 19, 20, 21, 22, 24, 27, 29, 30, 32, 35, 40, 45, 46, 47, 48	
Genetics and evolution	16, 23, 25, 26	
Microevolution and macroevolution	28, 31, 33, 34, 36, 37, 38, 39	
Human evolution	41, 42, 43, 44	

Classification; Viruses, Bacteria, and Protists; Fungi

Brief Yourself

Classification and Taxonomy

Organisms are scientifically named in Latin using a system devised by the 18th-century Swedish biologist, Carl Linnaeus, called the binomial system. Each type of organism is given a two-part name, consisting of the genus and species, as in *Homo sapiens* (humans) or *Canis domestica* (the domestic dog).

Taxonomy is the part of biology associated with assigning names to living things. Beyond the important task of ensuring each species has a unique, standardized name, taxonomists also attempt to classify organisms in ways which reflect their evolutionary relationships. A hierarchy of classification is used. Species are grouped into a genus (the plural is genera), genera are grouped into families, families into orders, etc. The hierarchy proceeds in the following order going from the smallest taxon (unit of classification) to the largest most inclusive. For example, humans are classified as:

species (*sapiens*) → genus (*Homo*) → family (Hominidae) → order (Primates) → class (Mammalia) → phylum (Chordata) →kingdom (Animalia).

Division is used in place of phylum for algae, plants, and fungi. Sometimes additional, modified units are used, such as subphylum or suborder to clarify relationships between organisms.

Diverse Microorganisms

Organisms considered here include the Prokaryotes (Archaebacteria and Eubacteria), photosynthetic and heterotrophic protists (protozoa, algae, some fungus-like organisms), fungi, and lichens. Viruses, particles which consist of a nucleic acid core surrounded by a protein coat, and which can reproduce within host cells, are also discussed.

Test Yourself

1. Organisms are given a two-part name in a format called the ________ system.

2. The modern system of naming organisms employs the Latin language, and was originally devised by the biologist _____________.
   a. Darwin
   b. Weismann
   c. Lamarck
   d. Linnaeus

3. In the technical name for the cat, *Felis* is the _________ and *domestica* is the __________.

4. List in order, from the largest group (most inclusive category) to the smallest:

   species, kingdom, family, order, genus, class, phylum

5. Fill in the blank.

Possible answers: parasite, protein, genome, retroviruses, capsid

(i) The viral _________consists of single- or double-stranded DNA or RNA.

(ii) The viral core of nucleic acid is surrounded by a _______ coat, or ________.

(iii) Some RNA viruses, called ____________, use a special enzyme (reverse transcriptase) to make more virus RNA in the host cell.

(iv) Viruses cannot reproduce outside of host cells, but only as _________ with a cell.

6. The disease AIDS is caused by the ______ virus.

7. Rod shaped bacteria are called __________; spherical bacteria are termed _________.

8. Bacteria lack a true __________, like other prokaryotes. They have a cell wall made of _____________ rather than the cellulose found in plant cells. Outside the cell wall is usually found a _________ or _______ layer.

9. _______________ bacteria obtain energy from sunlight and use carbon dioxide as a carbon source; _____________ bacteria use carbon dioxide and other inorganic substances for energy and carbon; __________ bacteria use sunlight for energy, but obtain carbon from organic compounds; and _____________ bacteria live on waste products and organic matter, or off of other organisms.

   Possible answers: photoheterotrophic, photoautotrophic, chemoheterotrophic, chemoautotrophic

Questions 10–13. True or false; if false, correct the statement to make it true.

10. Bacteria have organelles such as ribosomes, mitochondria, and chloroplasts.

11. Most common bacteria are Archaebacteria.

12. Archaebacteria have unique plasma membrane lipids, and unusual tRNA and rRNA.

13. *Rhizobium* is a eubacteria which ferments milk; *Lactobacillus* is an archaebacterium which fixes atmospheric oxygen, making it available for plants.

14. Bacteria can exchange genetic information in the form of __________; the rest of the bacterial genome is in the nucleoid region in the form of a _________ strand of DNA.

15. Bacteria reproduce by ___________ ___________ (two words).

16. True or false:
    (i) Protists may be unicellular or multicellular.
    (ii) Protists are heterotrophs.
    (iii) Protists have 9+2-type cilia or flagella.

17. All protists reproduce asexually by ____________. Some protists, like the ciliate *Paramecium*, can undergo __________ in

which ___________ are exchanged between cells.

18. Fission in protists is analogous in some respects to ______ in our cells.

19. __________ algae are very likely the ancestors of plants.

20. Red algae are usually found in ___________ (freshwater, salt water) habitats.

21. Kelp are a form of __________ algae.

22. __________ store food in the form of oil, which may help make them buoyant.

23. A toxic bloom of dinoflagellates, occurring in coastal waters, is called a ______ ________. (two words)

24. ____________ is an example of a protist which can be a either a heterotroph or photosynthetic autotroph, depending on environmental circumstances.

25. ____________ use cilia for motility, and can reproduce sexually by ___________.

26. Zoomastigotes include the trypanosomes which cause the diseases ________ _________ ___________ (three words) and _________ disease.

27. Amoeba and similar organisms are __________.

    a. rhizopods

    b. ciliates

    c. red algae

    d. zoomastigotes

28. True or false: Most bacteria are pathogenic (disease causing).

29. Methanogens and halophilic bacteria are examples of ______________.

30. The cell walls of fungi contain __________.

31. Myxomycota are the __________ __________ (two words) molds; whereas Acrasiomycota are the ___________ _________ (two words) molds.

32. Sporozoans cause parasitic diseases including ___________.

33. Many fungi have walls between cells in the hyphae called ___________, but these are often _________ (complete, incomplete) barriers.

34. Saprophytic fungi secrete __________ and absorb the digested material.

35. Most fungal cells are _________ (haploid, diploid).

36. Fungi which cause plant diseases include the __________ and _________.

37. Yeasts are members of ______________.

    a. ascomycetes

    b. basidiomycetes

    c. protists

    d. zygomycetes

38. Common organisms such as toadstools, mushrooms and shelf fungi are examples of _________.

    a. ascomycetes

    b. basidiomycetes

    c. protists

    d. zygomycetes

39. Fungal hyphae form matted structures called _____________.

40. Sexual reproduction idoes not occur in: ______________

    a. ascomycetes

    b. basidiomycetes

    c. deuteromycetes

    d. zygomycetes

41. True or false: Protists may include organisms which were the ancestors of the fungi, plants, and animals.

42. Lichen are a symbiotic association between a ___________ and a ___________.

43. The fungus in lichen are members of _____________.

    a. ascomycetes

    b. basidiomycetes

    c. protists

    d. zygomycetes

44. Fungal _________ penetrate algal cells in lichen, and pass on nutrients to the fungus.

45. Discuss the differences between RNA and DNA viruses.

46. Describe how bacteria may be grouped by staining characteristics.

47. Discuss the varied metabolic capabilities of bacteria.

48. Discuss the forms and habitats of photosynthetic protists.

49. Discuss the forms and habitats of heterotrophic protists.

50. Briefly characterize the fungi as a group.

51. What are some distinguishing points of zygomycetes?

Check Yourself

1. binomial **(Classification and taxonomy)**

2. d. Linnaus **(Classification and taxonomy)**

3. genus, species **(Classification and taxonomy)**

4. kingdom, phylum, class, order, family, genus, species. **(Classification and taxonomy)**

5. (i) genome

   (ii) protein, capsid

   (iii) retroviruses

   (iv) parasites

   Viruses defy easy classification. They are not alive, are not organisms, and can reproduce only as parasites within cells. A virus consists of a single- or double-stranded DNA or RNA core (the viral genome) surrounded by a protein coat (the capsid), and sometimes have a lipid-containing outer envelope as well. When a virus enters a host cell, it takes over the cell in order to make more viruses. Viruses are very specific, and can infect only one or a few types of host cells; a virus which infect one species generally can only be transmitted to a host of the same or closely-related species. Some of the best known viruses cause diseases in humans, ranging form the common cold (rhinoviruses) to AIDS (HIV, or human immunodeficiency virus), and Herpes (Herpes viruses). **(Viruses)**

6. HIV **(Viruses)**

7. bacilli, cocci. Bacteria are commonly described as having three basic shapes: rods (bacillus, bacilli), spheres (coccus, cocci) or corkscrew (spirillum, spirilla). Some form chains (such as streptococci) or clumps (staphylococci). Bacteria often have flagella for movement or pili (the singular is pilus) for attachment to surfaces. **(Bacteria—structure, classification, and reproduction)**

8. nucleus, peptidoglycan, capsule, slime. Bacterial cells are not subdivided into compartments by membranes, and do not contain organelles such as mitochondria or chloroplasts. They lack a true nucleus, although genetic material is confined to a central nucleoid region of the cell. Ribosomes are present, and are smaller than those found in eukaryotes. Bacteria possess cell walls made of polysaccharides cross-linked by peptides, making up a material called peptidoglycan. Around the cell wall may be a sort of capsule or slime layer, a glycocalyx made up of polypeptides, polysaccharides or a mixture of the two. **(Bacteria—structure, classification, and reproduction)**

9. photoautotrophic, chemoautotrophic, photoheterotrophic, chemoautotrophic. Bacteria are the most common prokaryotes and member of Kingdom Monera. They are ubiquitous, and play important roles in many natural processes and in some forms of human infectious disease. Bacteria are very small, single-celled organisms. They obtain energy from light or chemical compounds. They greatly vary in a number of respects; some need oxygen to live, others grow and reproduce only in the absence of free oxygen. Some bacteria can tolerate the high temperatures of hot springs and, in dormant spore form, withstand prolonged exposure to boiling water; other species die at the relatively low temperature of pasteurization (60 degrees C). **(Bacteria—structure, classification, and reproduction)**

10. False. Bacteria have organelles such as ribosomes. **(Bacteria—structure, classification, and reproduction)**

11. False. Most common bacteria are Eubacteria. Archaebacteria include methanogens (methane producers) found in swamps and animals intestines, and halophilic bacteria which live in saltwater, such as drying salt lakes or pools. These lack peptidoglycan cell walls, have unique metabolic pathways, possess plasma membranes with an unusual lipid makeup, and have unusual tRNA and rRNA sequences. Archaebacteria have been suggested as a separate kingdom, reflecting the unique history of these organisms. They diverged from the eubacteria early in the history of life. **(Bacteria—structure, classification, and reproduction)**

12. True **(Bacteria—structure, classification, and reproduction)**

13. False. *Lactobacillus* is a eubacteria which ferments milk; *Rhizobium* is a eubacterium which fixes atmospheric nitrogen, making it available for plants. Eubacteria (true bacteria) include most common bacteria. Nitrifying chemoautotrophs like *Nitrobacter* are important in recycling nitrogen. Photoautotophs (e.g., *Anabaena* and *Nostoc)* are found in all bodies of water. Photoheterotrophic bacteria metabolize organic compounds in soils and mud. Chemoheterotrophs include *Rhizobium*, which fixes atmospheric nitrogen and is associated with legumes and *Lactobacillus* which ferments milk into yogurt; and pathogens for humans, other animals, and plants (e.g., *Staphylococcus*, *Salmonella*, *Clostridium*). *Escherichia coli*, found in large number in the intestines of humans and other animals, makes vitamin K which is absorbed by our bodies. **(Bacteria—structure, classification, and reproduction)**

14. plasmids, circular **(Bacteria—structure, classification, and reproduction)**

15. binary fission. The bacterial genome, which is haploid, is located in the nucleoid region of the cell and consists of a single circle of DNA. Additional DNA is found in the periphery of the cell in the form of plasmids, small genetic units that can be exchanged between cells (the process of conjugation). Bacteria reproduce by binary fission, which is the process of one cell dividing into two. **(Bacteria—structure, classification, and reproduction)**

16. (i) True

    (ii) False. Some protists are heterotrophs, others can photosynthesize (algae, euglenoids, etc.)

(iii) False. Some but not all protists have cilia or flagella—which if present is always of the 9 + 2 type.

Kingdom Protista (Protoctista) is difficult to describe, as these eukaryotic organisms have been placed here because they were difficult to place anywhere else. Protista is a kingdom which bears a direct evolutionary relationship to every other kingdom. Typically, protists are single-celled, aquatic or marine, need free oxygen, possess 9 + 2- type flagella or cilia for motility. **(Protists—general, types and reproduction)**

17. fission, conjugation, micronuclei. Protists reproduce asexually by fission, analogous in some respects to mitosis in higher forms. Some protist can also reproduce sexually. For instance, ciliates like *Paramecium* can undergo conjugation: haploid micronuclei are exchanged between two cells, and fuse into a diploid macronucleus in each cell. The resulting diploid cells can divide further by asexual fission. **(Protists—general, types and reproduction)**
18. mitosis **(Protists—general, types and reproduction)**
19. green **(Protists—general, types and reproduction)**
20. salt water **(Protists—general, types and reproduction)**
21. brown **(Protists—general, types and reproduction)**
22. diatoms **(Protists—general, types and reproduction)**
23. red tide **(Protists—general, types and reproduction)**
24. *Euglena*. **(Protists—general, types and reproduction)**
25. a. ciliates (or *Paramecium*), conjugation. **(Protists—general, types and reproduction)**
26. African sleeping sickness, Chagas' **(Protists—general, types and reproduction)**
27. a. rhizopods **(Protists—general, types and reproduction)**
28. False. Most bacteria are either innocuous or beneficial to other organisms. **(Bacteria—structure, classification, and reproduction)**
29. Archaebacteria. **(Bacteria—structure, classification, and reproduction)**
30. chitin **(Fungi—major groups)**
31. plasmodial slime, cellular slime **(Protists—general, types and reproduction)**
32. malaria **(Protists—general, types and reproduction)**
33. septa, incomplete **(Fungal structure)**
34. enzymes **(Fungi—major groups)**
35. haploid. Fungi generally form filaments (hyphae), the cells of which are incompletely divided by walls (septa). A mat of hyphae is called mycelium; hyphae penetrate the material or organism on which they are growing. Fungal cells are usually haploid and divide mitotically; or if diploid cells exist, they become

haploid by meiosis. Sexual reproduction also occurs. Spores, resistant and dormant structures, may be formed by sexual or asexual means. **(Fungal structure)**

36. smuts, rusts **(Fungi—major groups)**

37. aascomycetes—cup fungi, truffles and yeasts. Reproductive structures, asci, fuse, undergo meiosis, and produce spores. Asexual production of spores is via conidia (made by mitosis at the end of special hyphae). Yeast commonly reproduce asexually by budding, which is similar to binary fission. **(Fungi—major groups)**

38. b. **(Fungi—major groups)**

39. mycelia (singular, mycelium). **(Fungal structure)**

40. c. deuterm;ycetes. The Deuteromycota (imperfect fungi) are fungi that sexual lifestages have not been found. The Ascomycetes, Basidiomycetes, and Zygomycetes all have sexual stages. **(Fungi—major groups)**

41. True, at least according to one scheme of how life evolved. **(Protists—general, types and reproduction)**

42. alga, fungus. Lichens are actually a community of sorts, a symbiosis between a fungus (ascomycetes) and an alga. The body of the lichen is a fungus, which contain photosynthetic algal cells. Fungal hyphae absorb nutrients from these cells. The fungus provides a protective matrix and strong attachment to rocks or other substrates, and the alga provides the food. **(Lichens)**

43. a. ascomycetes **(Lichens)**

44. hyphae. **(Lichens)**

45. While DNA viruses replicate themselves in a manner like that used to make DNA in the host cell, RNA viruses are more complicated. RNA viruses may use a reverse transcriptase enzyme (retroviruses) or RNA replicase to make more RNA using the genomic RNA as a template. RNA is used to make DNA, which is then used to make more RNA for new virus particles. **(Viruses)**

46. Bacteria may be described by staining characteristics. Bacteria that retain the stain during the Gram staining process are called Gram-positive. Bacteria that do not retain the stain are termed Gram-negative. **(Bacteria—structure, classification, and reproduction)**

47. Bacteria may be (1) photoautotrophs (obtain energy from sunlight and use carbon dioxide as a carbon source), (2) chemoautotrophs (use carbon dioxide and other inorganic substances for energy and carbon), (3) photoheterotrophs (use sunlight for energy, but obtain carbon from organic compounds), or (4) chemoheterotrophs live on waste products and organic matter, or off of other organisms. **(Bacteria—structure, classification, and reproduction)**

48. This a diverse group, and it appears that members of the various phyla independently acquired chloroplasts (endosymbiotic theory). These include: green algae (mostly multicellular, probably ancestors of plants; cell walls partly or wholly cellulose, and food is stored as starch; mostly aquatic; includes *Spirogyra*, *Volvox*, and *Acetabularia*), red algae (multicellular; cell walls of cellulose, store food in a starch that is like glycogen, have chlorophylls *a* and *d*; mainly marine; uniquely lack flagellated cells at any point in life cycle; includes may seaweeds and source of agar), brown algae (multicellular; have chlorophylls *a* and *c*; most common seaweeds, including kelp), golden algae (have carotenoid and chlorophyll pigments; flagellated; have some silica in cell walls and store food in a unique carbohydrate

compound; an aquatic plankton), diatoms (marine and aquatic; shells of silica and organic material; store food as oil), dinoflagellates (cellulose cell wall; generally unicellular phytoplankton, with two flagella; some form toxic "red tides" in coastal waters; some are cellular symbionts of corals), and euglenoids (unicellular, flagellated, with no cell wall; a common form is *Euglena* which can ingest food or photosynthesize depending on environmental conditions). **(Protists—general, types and reproduction)**

49. These include the ciliates (have cilia for motility, and sometimes to sweep food into cell; can reproduce by conjugation and have micro- and macronuclei; *Paramecium* and *Tetrahymena* are noted ciliates), zoomastigotes (unicellular or colonial; one or more flagella; includes trypanosomes causing African sleeping sickness and Chagas' disease, as well as cellulase-producing, symbiotic inhabitants of the guts of termites), rhizopods (amoeboid organisms that move by pseudopodia; can cause intestinal diseases in animals and humans), sporozoans (amoeba-like parasites, causes diseases such as malaria), foraminiferans (marine, with shells of calcium carbonate and organic materials). also included in this group are the slime molds. Phylum Myxomycota consists of plasmodial slime molds, commonly seen as colored blotches on fallen, decaying trees—they form a large, multinucleate mass termed a plasmodium. Phylum Acrasiomycota, the cellular slime molds, are very different and quite unrelated to the plasmodial slime molds: they spend most of their life cycle as single amoeboid cells, congregating and forming fruiting structures to make spores when conditions are unfavorable. **(Protists—general, types and reproduction)**

50. Most fungi are multicellular or multinucleate, and non-motile. Saprophytic or parasitic, they obtain food by secreting enzymes onto a substrate, and absorbing the digested material into their cell, or by feeding directly off a host organism. Cell walls are typically made up of chitin and other substances. **(Fungi—major groups)**

51. In Zygomycetes the hyphae lack cell walls except during reproduction. Reproductive structures (zygosporangia) are formed by the union of gametangia; spores are also asexually produced. **(Fungi—major groups)**

Grade Yourself

Circle the question numbers that you had incorrect. Then indicate the number of questions you missed. If you answered more than three questions incorrectly, you will then have to focus on that topic. (If a topic has less than three questions and you had at least one wrong, we suggest you study that topic also. Read your textbook, a review book, or ask your teacher for help.)

Subject: Classification; Viruses, Bacteria, and Protists; Fungi.

Topic	Question Numbers	Number Incorrect
Classification and taxonomy	1, 2, 3, 4	
Viruses	5, 6, 45	
Bacteria—structure, classification, and reproduction	7, 8, 9, 10, 11, 12, 13, 14, 15, 28, 29, 46, 47	
Protists—general, types and reproduction	16, 17, 18, 19, 20, 21, 22, 23, 24, 25, 26, 27, 31, 32, 41, 48, 49	
Fungal structure	33, 35, 39	
Fungi—major groups	30, 34, 36, 37, 38, 40, 50, 51	
Lichens	42, 43, 44	

Survey of the Animal Kingdoms

Brief Yourself

General Characteristics of Animals

Animals are organisms that are almost all motile, lack a cell wall, are heterotrophs, develop through a series of embryonic stages, are multicellular and have cell and tissue differentiation. The predominant form is diploid, with brief haploid stages in the form of gametes. Sexual and asexual reproduction are practiced, sometimes both in some forms, although sexual is the most common form. Animals generally have large, non-motile eggs and small, flagellated sperm. Fertilization may be external or internal.

Animals may be classified on these bases among others: presence or absence of a true body cavity or coelom, segmentation of the body, developmental patterns, exo- or endoskeletons, body symmetry, and development of a head region.

The Major Animal Phyla

There are nine major phyla in Subkingdom Metazoa: Chordata, Arthropoda, Annelida, Mollusca, Nematoda, Coelenterata, Echinodermata, Porifera, and Platyhelminthes. Each has its own particular characteristics, and a range of diversity within it. There are also rare animal phyla.

Basic Body Plans (Bilateral and Radial)

Metazoan animals have two basic body types. The bilateral body has only one plane of symmetry, dividing the body into right and left halves. The radial symmetry plan involves a disk-like or cylindrical body, with unlimited plans of symmetry. For example, the sea urchin has a five part body plan.

Test Yourself

1. Animals (pick all that apply):
   a. are multicellular
   b. are autotrophic
   c. are heterotrophic
   d. are motile
   e. are sessile
   f. have cell walls
   g. have no cell walls
   h. develop through embryonic stages
2. True or false: Most animals and species are invertebrate.
3. True or false: Animals have a predominantly haploid life cycle.
4. Evolutionary modifications in animals body plan, and the basis for classification involve which of the following (pick all that apply):
   a. body symmetry
   b. a true body cavity (coelom)
   c. segmentation
   d. development of a cephalic (head) end
   e. pattern of embryonic development
5. The only haploid stage in the animal life cycle is in the form of a ___________.
6. One group of animals is notably distinct and classified apart form all others:
   a. insects
   b. Arthropoda
   c. Cnidaria
   d. Porifera
   e. Chordata (includes vertebrates)

Questions 7–16. Match the major animal phylum with the examples in the list.

Possible examples: (i) jellyfishes and hydra, (ii) hookworms and other round worms, (iii) earthworms, (iv) sponges, (v) crustaceans and insects, (vi) ribbon worms, (vii) fishes and birds, (viii) seastars (starfishes) and sea urchins, (ix) flatworms including planarians, (x) marine comb jellies, (xi) probiscus worms (xii) snails and clams.

Phylum

7. Porifera
8. Mollusca
9. Annelidia
10. Echinodermata
11. Chordata
12. Cnidaria (Coelenterata)
13. Platyhelminthes
14. Nematoda
15. Ctenophora
16. Nemertea (Nemertina)

Questions 17–23. Answer with numbers 7–16 (the same number as used in the questions above which bear the numbers for the names of the phyla—i.e. if the answer is Porifera, write "7.") More than one answer may be used. Write "all" if the statement applies to all the listed phyla.

17. Which of the above phyla are radially symmetrical?
18. Which of the above phyla are bilaterally symmetrical?
19. Which of the above phyla originated from a common protozoan ancestor according to current knowledge?
20. Which phyla consist of protostomates?
21. Which phyla consist of deuterostomates?
22. Which phyla are coelomates?

23. Which phyla are acoelomates?

24. A true body cavity is called a __________. Organs found in this body cavity in humans and other mammals include ___________, ___________, and _____________.

25. Give three examples of segmented animals.

26. Compare a seastar (starfish) and a crayfish with regard to body symmetry.

27. Animal embryos consist of three germ layers. Name them.

28. How is a false coelom different from a true coelom?

29. Unlike most other animals, sponges are ___________ (pick all that apply):

    a. sessile

    b. heterotrophic

    c. photosynthetic autotrophs

    d. made up of cells with a cell wall to give the sponge shape

30. Sponges consist of two layers of tissue sandwiching a jellylike layer called _________. ____________ are specialized cells which beat their flagella to move water through the sponge. _______________ course through the jelly-like layer, pick up food and lay down glassy fibers called _____________. Sponges have a _________ form which is free-swimming. Sponges lack ____ _____ tissues and thus have no nervous system.

31. A single opening serves as both a mouth and an anus for ____________ , once called coelenterates. Digestion is carried out in the ____________ cavity. Cnidarians are unique in that they are the only animals with stinging cells, or _____________ , used to immobilize prey and ward off predators. Jellyfishes show one form of organization, the cuplike __________ . Jellyfishes use ______ as a hydrostatic skeleton to support their shape. Hydra exemplify the other form, the _________, which has a mouth at one end circumscribed by a set of feeding ____________ . Cnidarians have nerve, contractile and sensory cells that let them monitor and respond to their environment. Colonial forms exist such as *Physalia*, popularly known as the ____________ .

32. Flat worms also have a _______________ cavity with a single mouth/anus. Flatworms for the most part are sexual and some species are ________________ , with both _________ and __________ gonads. Mating involves the simultaneous exchange of sperm. Tapeworms are a major human and animal parasite, attaching to the gut wall by means of a hooked __________ .

33. Protostomes differ from deuterostomes in that (pick all that apply):

    a. Protostome cleavages in the embryo are radial, and at right angles to the main embryonic axis, whereas deuterostomes cleave obliquely in a spiral fashion.

    b. In deuterostomes, the first opening in the embryo becomes the mouth, the second the anus.

    c. In protostomes, the first opening in the embryo becomes the mouth, the second (or another) the anus.

    d. The coelom in protostomes starts from folds in the gut.

34. Gastropods include molluscs such as __________ and the _________ or sea slugs.

    a. chitons

    b. sea stars

    c. snails

    d. nudibranchs

35. Most molluscs have soft-bodies and protected by a rigid shell. Characteristic features are the _________, draped around the body and which secretes shells in those molluscs which have them; and a muscular foot (easily seen on a open clam or a snail). A rough tongue, the _________ is found in some, and all molluscs other than garden snails have distinct sexes. Garden snails, by contrast are ___________ .

36. Annelids include the familiar ____________. _______ act as kidneys and the nerve cords bunch up in masses of cells called ______

which control nearby functions. Earthworms have a ___________ with blood pumped by muscular vessels.

37. True or false; if false, correct to make the statement true. Annelids are bilaterally organized, segmented, without a true coelom.

38. The most advanced of the molluscs are the ____________ including the octopus and cuttlefish. These animals are unique among molluscs in that they possess a __________ ____________ ___________ (three words). Tentacles take the place of the muscular foot, have highly-developed __________ system and complex ___________, similar in some respects to those found in humans.

39. ____________ are the most diverse and one of the most plentiful kinds of animals. Characteristics include jointed _________________ (giving them their phylum name) used for defense, locomotion, sensing, and feeding; a hard chitinous ______________ exoskeleton (which forces them to grow by molting); ____ circulatory system; __________ or gills for breathing (book lungs in spiders) ; specialized nerve cells such as eyes; and sometimes complex social structures involving division of __________ in colonies.

40. Spiders, while they are _________, are not ___________.

41. Chilopoda and diplopoda are the ___________ and ___________, respectively. The former are ___________ while the latter feed on detritus.

42. Insect have ___________ Malpigian tubules which function as kidneys.

    a. Malpighian tubules

    b. nephridia

    c. renal glands

    d. nephrons

43. Insects have _________ of legs, one set of _______ , one or two pairs of _________ and a body formed of a head, ________ and abdomen. The head often exhibits specialized _____________, as in grasshoppers.

44. Fertilization occurs ____________ (internally, externally) in insects.

45. The change form larva to pupa to adult is called _____________ and is controlled by _____________.

46. Echinoderms possess an ___________, and a water _________ ________ (two words)for movement, attachment, and feeding. In sea stars the_______ feet are readily observed parts of this system. Alone among the ___________, adult echinoderms have a radial body arrangement.

47. Chordates include both ______________ and _____________ animals. All chordates possess a _______________ , which supports the body as a rod, a dorsal __________ __________ (two words), a ___________ , and ______ slits at some time in the life of the animal. At some point in development or adult life, all chordates also have a tail that extends beyond the ________.

48. Vertebrates are chordates with a bony _________ and ________ protecting the _________ ___________ ___________ (three words).

49. Match each example with the appropriate vertebrate class in the list.

Examples:

(i) lampreys and other jawless fishes with cartilaginous skeletons. Possess a two-chambered heart.

(ii) endothermic, one ovary in females, light bones, oviparous with a four-chambered heart and special lung arrangement.

(iii) Posses mammary glands and hair, endothermic, have a muscular sheet dividing the thorax from the abdomen, and no nuclei in adult red blood cells. Possess a four-chambered heart.

(iv) Shelled eggs with fluid-filled amnion, ectothermic, internal fertilization even in oviparous forms, terrestrial aquatic and marine.

(v) Cartilaginous skeleton, includes rays, digestive, reproductive and urinary systems exit the body via a cloaca, often have special senses allowing them to detect the movement of prey in the water. Posses a two-chambered heart.

(vi) Bony fishes, gills cover by a operculum, water is pumped over the gills so that breathing does not require forward body movement has it does in sharks. Possess a two chambered heart.

(vii) Have larval and adult forms, lay eggs, generally live near water in order to reproduce, have wet skin that assists in gas exchange.

Classes of vertebrates:

a. Agnatha

b. Chondrichthyes

c. Osteichthyes

d. Amphibia

e. Reptilia

f. Aves

g. Mammalia

50. The lower vertebrates include the classes (give letters from question 49 above), most of which have two or three chambered hearts. _______________ are an exception and have a four-chambered heart.

51. The vertebrate may reproduce through either ______ or ________ fertilization.

52. Unique features of birds are ___________ and air-filled __________, and a __________ for grinding food.

53. Modern birds are thought to have perhaps originated from the now-extinct _______________.

54. Mammals are unique in that they have __________, and secrete__________ for their young,

55. __________ are egg-laying mammals. ______________are the __________ mammals. All other mammals are termed eutherian, that is the ______________ mammals. Humans are ____________.

56. What features demonstrate that a porpoise is a mammal?

57. Amphibians undergo _______________ in which a ________ larva with a ______tail changes into an adult with lungs. This process is controlled by the __________ gland. Amphibian eggs are not constructed to resist ______, and thus must be laid in ____________. Amphibians as a result spend their lives in or near ________.

Check Yourself

1. a, c, d, g, h are true **(General animal characteristics)**

2. True **(General animal characteristics)**

3. False. Animals have a predominantly diploid life cycle. **(General animal characteristics)**

4. All are correct. **(General animal characteristics)**

5. gamete (sex cell) **(General animal characteristics)**

6. d. Porifera diverged from other animal groups very early in evolutionary history, and thus are only distantly related to other groups of animals. **(General animal characteristics)**

7. iv. sponge **(Major animal phyla)**

8. xii. snails and clams **(Major animal phyla)**

9. iii. earthworms **(Major animal phyla)**

10. viii.seastars **(Major animal phyla)**

11. vii. fishes and birds **(Major animal phyla)**

12. i. jellyfishes and hydra **(Major animal phyla)**

13. ix. flatworms including planarians **(Major animal phyla)**

14. ii. hookworms and other round worms **(Major animal phyla)**

15. x. marine comb jellies **(Major animal phyla)**

16. xi. probiscus worms **(Major animal phyla)**

17. 12 cridaria and 15 ctenophora **(Basic body plans)**

18. All but 7, 12 and 15 are bilaterally symmetrical (Echinoderms have radial organization with bilateral characteristics and often radial adult forms produce bilateral larvae.) **(Basic body plans)**

19. All. Protists or protoctists are considered an ancestral group for all these phyla. **(Basic body plans)**

20. 8 mollusca, and 9 annelidia **(Basic body plans)**

21. 10 echinodermata, and 11 chordata **(Basic body plans)**

22. 8 mollusca, 9 annelidia, 10 echinodermata, and 11 chordata **(Basic body plans)**

23. 13 platyhelminthes, and 16 nemertea **(Basic body plans)**

24. coelom (pronounced "seal-um"), heart, lungs, kidneys or other abdominal or thoracic organs. **(Basic body plans)**

25. earthworms, crayfish, insects. Vertebrates could be another, less obvious choice. **(Basic body plans)**

26. A seastar is radially organized (as around a circle) but with some bilateral characteristics, whereas a crayfish is clearly bilaterally arranged (as are humans). **(Lower invertebrates)**

27. Ectoderm, mesoderm, and endoderm. **(General animal characteristics)**

28. A false coelom lacks a lining called the peritoneum. **(Basic body plans)**

29. a. sessile. All animals are heterotrophs, but no animals are photosynthetic autotrophs or have cell walls. **(Lower invertebrates)**

30. mesoalea, choanocytes, amoebocytes, spicules, larval, true **(Lower invertebrates)**

31. cnidarians, gastrovascular, nematocysts, medusa, water, polyp, tentacles, Portuguese man-of-war **(Lower invertebrates)**

32. gastroventricular, hermaphroditic, male, female, scolex **(Lower invertebrates)**

33. c. snails **(Protostomates: Molluscs and Annelids)**

34. c snails, d nudibranchs **(Protostomates: Molluscs and Annelids)**

35. mantle, radula, hermaphrodites **(Protostomates: Molluscs and Annelids)**

36. earthworms, Nephridia, ganglia, closed circulation **(Protostomates: Molluscs and Annelids)**

37. change to read, ". . .with a true coelom" **(Protostomates: Molluscs and Annelids)**

38. cephalopods, closed circulatory system, nervous, eyes **(Protostomates: Molluscs and Annelids)**

39. Arthropods. appendages, exoskeleton, open, trachea, labor **(Protostomates: Arthropods)**

40. arthropods, insects **(Protostomates: Arthropods)**

41. centipedes, millipedes, carnivorous **(Protostomates: Arthropods)**

42. a. Malpighian tubules **(Protostomates: Arthropods)**

43. three pairs, antennae, wings, thorax, abdomen, appendages **(Protostomates: Arthropods)**

44. internally **(Protostomates: Arthropods)**

45. metamorphosis, hormones **(Protostomates: Arthropods)**

46. endoskeleton, vascular system, tube, coelomates **(Deuterostomates: Echinoderms)**

47. invertebrate, vertebrates, notochord, nerve cord, pharynx, gill, anus **(Deuterostomes: Chordates)**

48. skull, spinal column, central nervous system **(Deuterostomes: Chordates)**

49. a. i, b. v, c. vi, d. vii, e. iv, f. ii, g. iii **(Deuterostomes: Chordates)**

50. a, b, c, d, and e, crocodiles **(Lower vertebrates)**

51. internal or external **(Lower vertebrates)**

52. feathers, bones, gizzard **(Birds)**

53. dinosaurs **(Birds)**

54. hair, milk **(Mammals)**

55. Monotremes, marsupials, pouched, placental, eutherians **(Mammals)**

56. Porpoises have a few hairs on their bodies, mature red blood cells without nuclei, seven cervical vertebrae, and produce milk and suckle their young Despite misleading external appearances, porposes are clearly mammals. **(Mammals)**

57. metamorphosis, gilled, tail, thyroid, drying, water, water **(Lower vertebrates)**

Grade Yourself

Circle the question numbers that you had incorrect. Then indicate the number of questions you missed. If you answered more than three questions incorrectly, you will then have to focus on that topic. (If a topic has less than three questions and you had at least one wrong, we suggest you study that topic also. Read your textbook, a review book, or ask your teacher for help.)

Subject: Survey of the Animal Kingdoms

Topic	Question Numbers	Number Incorrect
General animal characteristics	1, 2, 3, 4, 5, 6, 27	
Major animal phyla	7, 8, 9, 10, 11, 12, 13, 14, 15, 16	
Basic body plans	17, 18, 19, 20, 21, 22, 23, 24, 25, 28	
Lower invertebrates	26, 29, 30, 31, 32	
Protostomates: Molluscs and Annelids	33, 34, 35, 36, 37, 38	
Protostomates: Arthropods	39, 40, 41, 42, 43, 44, 45	
Deuterostomates: Echinoderms	46	
Deuterostomates: Chordates	47, 48, 49	
Lower vertebrates	50, 51, 57	
Birds	52, 53	
Mammals	54, 55, 56	

Homeostasis, Respiration, and Circulation

Brief Yourself

Body Systems

The major systems of the human body serve to provide a person with all the functions necessary for life—obtaining and assimilating food, eliminating wastes, regulating behavior and physiology, defending against infection, obtaining information about the environment, etc.

Circulatory Systems

Circulatory systems are of two main types: open and closed. In closed systems (as in mammals), the blood is contained at all times in the heart of vessels. In open systems, as are found in insects, the blood leaves arteries, fills spaces termed sinuses, and is returned to veins.

The Heart

The primary pump of the circulatory system is the heart (some animals have more than one), which assists in moving blood by the progressive contraction of the muscles in the walls of the arteries. Blood pressure falls in capillaries and veins. Due to low venous pressure, a major agent for the return of blood to the heart is muscular action in the limbs, which squeeze blood back toward the heart. The backward flow of blood is prevented by valves in the veins.

Blood

Blood is one of the most complex tissues in the body. It carries oxygen and carbon dioxide, hormones, water, body heat, and nutrients. The red cells carry the gases, partly complexed with hemoglobin. Other cells, such as white cells function in immunity, while components such as fibrinogen and platelets provide for the clotting of blood to halt bleeding from injuries. Blood is mainly water; the non-cellular portion of the blood is called plasma.

Respiration Systems

Respiratory systems come in various forms, but all promote the exchange of gases within an organism. Lungs and gills are two common forms of moving gases. Smaller organisms don't need respiratory systems—simple diffusion suffices to move carbon dioxide out and oxygen in.

Homeostasis

In order to live, an organism must keep the internal environment stable within a certain range of values (for pH, glucose, oxygen, and carbon dioxide concentrations, temperature, etc.). This effort to maintain the constant internal environment is called homeostasis.

Test Yourself

1. Name the human body system associated with each function:
    a. defense against infectious bacteria, fungi, and viruses
    b. movement
    c. controls motor functions and coordinates with sensory information
    d. processes food so it can be assimilated
    e. eliminates nitrogen-containing (nitrogenous) wastes
    f. gas exchange with the environment
    g. moves nutrients, cells, gases throughout the body
    h. production of children
2. Describe the path red blood cells take through a mammal's body going from the left atrium back to the left atrium.

Questions 3–5. Matching.

Possible answers: fish, amphibians and reptiles, birds and mammals, veins.

3. Which animals have a four-chambered heart?
4. Which animals have a three-chambered heart (two atria, one ventricle)?
5. Which animals have a two-chambered heart?

Questions 6–11. Matching.

Possible answers: closed circulation, open circulation, systole, diastole, ventricle, atrium, arteries, veins.

6. Muscular vessels that carry blood away from the heart are called ________.
7. The pumping chamber of heart.
8. Arteries and veins joined by capillaries in this type of circulatory system.
9. Arteries and veins not joined by capillaries in this type of circulatory system.
10. Period when ventricles contract.
11. Period when ventricles relax and receive blood from atria.
12. The hydra survives without a special transport system since
    a. it possesses a nerve net
    b. it can ingest food with the aid of tentacles containing stinging cell
    c. most of its cells are in direct contact with a watery environment
    d. all of its cell can live without using oxygen
13.

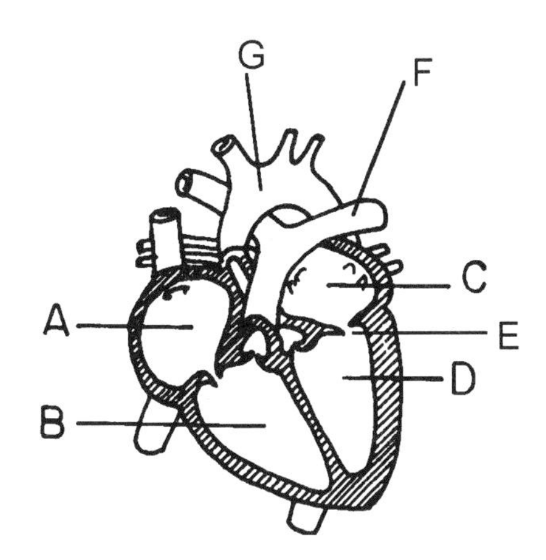

   a. A valve that keeps blood from flowing back into an atrium is found at letter ____
   b. The chamber which pumps blood to the body is indicated by letter ______
   c. Which structures contain oxygenated blood?
   d. Which vessel takes blood to the lungs?
   e. Which structures are ventricles?
   f. Which structures are atria?

14. In the mammalian circulatory system, most of the end products of nutrition are carried in the

    a. white blood cells

    b. erythrocytes

    c. platelets

    d. plasma

Questions 15–21. Matching.

Possible answers: dissolved ions, tricuspid, bicuspid (mitral), proteins, serum, plasma, blood, cardiac output, semilunar valve. (Not all answers are used.)

15. Atrioventricular valve on the right side of the heart.

16. Atrioventricular valve on the left side of the heart.

17. At junction of left ventricle and aorta.

18. Volume of blood pumped by the left ventricle.

19. Electrolytes.

20. Helps control (buffer) blood pH.

21. Is made up of water, dissolved substances, proteins, but no cells.

22. Who is credited with discovering the circulation of the blood, and in what century and country did he live?

23. True or false: The cells of the heart muscle are different from smooth or skeletal muscle cells.

24. True or false: Cardiac muscle cells do not have an intrinsic ability to contract.

25. The pacemaker of the heart is located in the right atrium and is called the ________ ___________ (two words, do not use abbreviations).

26. Impulses from the _____________ nervous system speed up the heart rate by affecting the SA node; impulses from the ____________ nervous system slow down the heart rate.

27. The major artery of the mammalian body is the ___________, the major vein, the ___________ ________ (two words). The heart's own circulation is the ____________ circulation, which branches off from the ________.

28. In a closed circulatory system, blood is contained within __________; in an open circulatory system, the blood freely mixes with the fluid bathing cells. In the latter type of system, blood often collects in open areas called __________.

29. Why doesn't a hydra or planarian flatworm need a circulatory system?

30. Give an example of an animal with more than one heart.

31. Blood carries ________ in many animals, but not in insects.

32. What anatomical advantage does the heart of a crocodile have over the hearts of other reptiles?

33. How is the circulation arranged in an amphibian like a frog?

34. Blood in fish travels through two capillary beds before it returns to the heart to be recirculated. Describe them.

35. Where does the exchange of nutrients and gases occur in the tissues in mammals?

36. In the human heart,

    (i) which chamber has the thickest muscular walls?

    (ii) which chamber contains the most highly oxygenated blood?

    (iii) which chamber contains the least oxygenated blood?

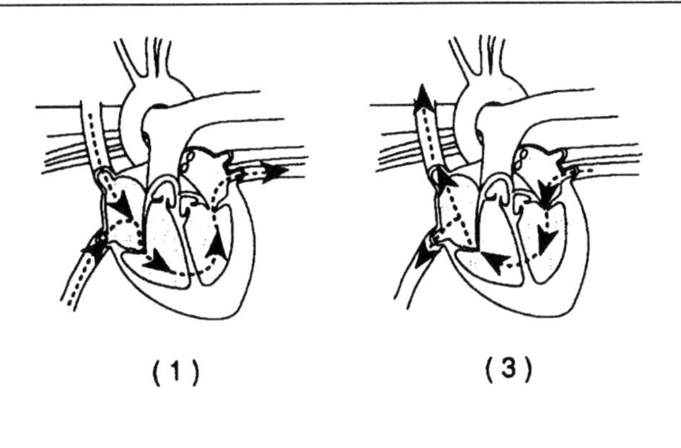

(1) (3)

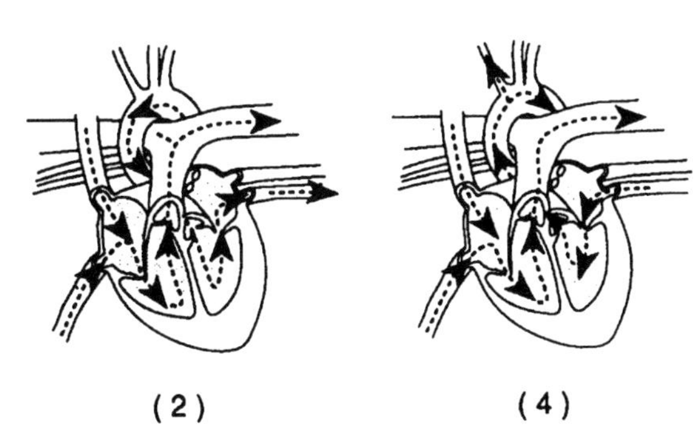

(2) (4)

(iv) which of these four pictures correctly shows how blood is pumped in the heart?

37. How is blood moved through the circulatory system of a vertebrate?

38. How do veins and arteries differ in terms of structure?

39. How is (i) oxygen and (ii) carbon dioxide carried in the blood in mammals?

40. Describe an erythrocyte.

41. Describe the respiratory systems of

(i) a fish

(ii) a frog

(iii) a person

(iv) a bird

42. How does air compare to water as a medium for carrying gases? Mark these statements as true or false.

(i) Air contains less oxygen than water.

(ii) Cold air has more oxygen than warm air.

(iii) Cold water can contain more oxygen than warm water.

(iv) Water contains less oxygen than air.

43. How do insects breathe? Pick all statements that are true.

a. Gases are pumped in and out using lungs.

b. Gases enter and leave through openings in the abdomen called spiracles.

c. Gases move into the insect only by diffusion.

d. Gases are pumped in and out of an insect through muscular movement.

44. The smallest sacs in the mammalian lung, where gas exchange occurs are called ________.

45. How do gases move from the lung into the blood and visa versa?

a. Active transport

b. Diffusion

c. Active transport and diffusion

(iv) Bulk flow

46. The respiratory center of the brain is the ________________.

47. How do spiders breathe?

48. What is meant by the partial pressure of a gas?

49. How does blood clot?

a. Blood dries, in forming a clot.

b. Fibrinogen is converted to fibrin in a series of reactions, and the fibrin holds back cells.

c. Fibrin is converted to fibrinogen in a series of reactions, and the fibrinogen holds back cells.

50. Describe the clotting of blood.

51. How is air moved in and out of the chest of a person? Of a frog?

52. How is the rate of breathing controlled—by the concentration of oxygen or carbon dioxide in the blood?

53. How is the surface area of lungs and gills important in respiration? How is the surface area increased?

54. What are some functions of the blood and circulatory system?

55. What is homeostasis?

    a. maintaining a constant internal environment.

    b. eating a balanced diet

    c. maintaining a constant body temperature

56. Give an example of homeostasis.

57. The concept of homeostasis was introduced by ___________ in the 19th century, and the term was coined by ___________ in the 1930s.

    a. Charles Darwin

    b. Walter B. Cannon

    d. Meselson and Stahl

    e. Claude Bernard

    f. August Weissman

58. What are some conditions maintained in homeostasis in mammals?

59. What is lymph? What are the functions of the lymphatic system?

60. How and where does lymph return to the circulatory system in mammals?

Check Yourself

1. a. immune

   b. musculo-skeletal

   c. nervous

   d. digestive or gastrointestinal

   e. excretory, urogenital, or urinary

   f. respiratory

   g. circulatory

   h. reproductive **(Body systems)**

2. Leaves left atrium - left ventricle - aorta - arteries - arterioles - capillaries - venules - veins - vena cava - right atrium -right ventricle - pulmonary artery - lungs - pulmonary vein - left atrium **(Circulatory systems)**

3. birds and mammals **(The heart)**

4. amphibians and reptiles **(The heart)**

5. fish **(The heart)**

6. arteries **(Circulatory systems)**

7. ventricle **(Circulatory systems)**

8. closed circulation **(Circulatory systems)**

9. open circulation **(Circulatory systems)**

10. systole **(The heart)**

11. diastole **(The heart)**

12. c. The limited size of the hydra allows its cells to obtain oxygen and nutrients without a circulatory system. **(Circulatory systems)**

13. a. E

    b. D

    c. C, D, G

    d. F

    e. B, D

    f. A, C **(Circulatory systems)**

14. d. plasma **(Blood)**

15. the tricuspid. The function is suggested by the name atrioventricular. **(The heart)**

16. bicuspid (or mitral valve) **(The heart)**

17. semilunar valve. Another semilunar valve is located at the junction of the right ventricle and the pulmonary artery. Semilunar valves prevent backflow of blood at the point where the blood leaves the heart. **(The heart)**

18. cardiac output. At rest, an adult human heart will pump about 5 liters of blood per minute. **(The heart)**

19. dissolved ions (such as calcium, sodium, potassium). **(Blood)**

20. Proteins. Also, other compounds such as the bicarbonate ion, help buffer the blood (maintain its pH). **(Blood)**

21. Plasma. The liquid part that remains after cellular elements are removed. Serum is what remains after clotting elements have been removed from the blood. **(Blood)**

22. William Harvey, and English physician, described the circulation of the blood in 1628 (17th century). **(Circulatory systems)**

23. True. The cardiac muscle of the heart is distinct in structure and physiology from skeletal and smooth muscle. Cardiac muscle cells have an intrinsic capacity to contract without outside stimulation. Cardiac muscle fiber appears different from other muscle fiber under the microscope. **(The heart)**

24. False **(The heart)**

25. sinoatrial node. The heartbeat is controlled at the sinoatrial node in the right atrium. Hormones, exercise and various medications can all affect this control center and thus the heart rate. **(Control of circulation)**

26. sympathetic, parasympathetic **(Control of circulation)**

27. aorta, vena cava, coronary, aorta **(The heart)**

28. vessels, sinuses **(Circulatory systems)**

29. The small size of hydrazoans and planarians means they have an adequate ratio of surface area to volume of tissues, and therefore simple diffusion can take care of requirements for the exchange of materials such as gases, nutrients, and water. **(Circulatory systems)**

30. An earthworm **(Circulatory systems)**

31. oxygen **(Blood)**

32. A crocodile has a four-chambered heart, like a mammal; blood from the ventricles does not mix as it does in amphibians and reptiles with two atria but only one ventricle. In some reptiles, such as lizards, a septum (wall) partially divides the ventricle and reduces the mixing of oxygenated and deoxygenated blood. Amphibians have a ridge of tissue that is less effective at dividing the single ventricle, but which nonetheless helps reduce mixing of blood. **(The heart)**

33. Frogs pump blood to the lungs and the rest of the body from a single ventricle. There are pulmonary (lung) and systemic (body) circuits in the circulation. Oxygen is picked up in the lungs and from the skin. **(Circulatory systems)**

34. Fish, with a two-chambered heart, pump blood through one circuit, in which the blood goes through the gills, picking up oxygen, and then moves on through the body before returning to the heart. This is in contrast to two-circuit circulation, in which blood is pumped from the heart into the lungs, and returns to heart before being pumped out into the rest of the body. **(Circulatory systems)**

35. Gases are exchanged in the lungs (in the alveoli) and in the capillary beds in the tissues. Oxygen diffuses from a region of higher concentration (the alveolus) to a region of lower concentration (the blood). Carbon dioxide moves in the lung in the opposite direction, from a region of higher concentration (the blood) to a region of lower concentration (the alveolus). In the tissues, carbon dioxide moves from the tissues to the blood. **(Respiration, gas exchange, control)**

    Nutrients are picked up in certain tissues (glucose in the liver) and released in other tissues, by diffusion or active transport out of capillaries. **(Blood)**

36. (i) The left ventricle, which pumps to the entire body.

    (ii) Blood in the left atrium and ventricle is the most highly oxygenated.

    (iii) Blood in the right atrium and ventricle contains the least amount of oxygen.

    (iv) Illustration 2 is correct. Examine the others to see they make no sense. **(The heart)**

37. Blood is pumped out of the heart by the contraction of the ventricle, and enters the arteries, where it is further pushed along by muscles in the arterial walls. Blood pressure drops in the capillaries and veins. Venous blood is returned partly by the action of muscles used in body movement. Some of the fluid returns from the tissue as lymph. Note that the capacity of the arteries and especially the veins to hold blood can be regulated, for example, in vasodilation the veins increase in size by relaxation of the muscles in the vessels, and more blood is retained in this part of the circulatory system. When blood is needed to

combat blood loss, the veins can be tightened (vasoconstriction) to force blood into other parts of the system. Additional blood is also stored in the spleen, allowing the animal to, in effect, give itself a transfusion of blood to make up for excess blood loss, as might result from an injury. **(Circulatory systems)**

38. Arteries have thick, elastic walls, made up of layers of connective tissue, elastic fibers, and smooth muscle. They continue the heart's work by contracting and squeezing blood along producing a pulse. Veins are thinner-walled vessels, generally larger than corresponding arteries, and expand more readily than do the arteries. Both arteries and veins have an inner lining of endothelium cells; these cells constitute the single-cell wall of capillaries. Finally, veins have valves, arteries do not. **(Circulatory systems)**

39. (i) Ninety-five percent of oxygen is carried by hemoglobin in the erythrocytes (red blood cells or RBC). Only about 5 percent is dissolved in the blood plasma. In regions of high oxygen content such as the lungs, oxygen is readily taken up by hemoglobin. In regions of low oxygen content such as the tissues, oxygen is given up.

    The relationship between oxygen in the surroundings is not a simple one. If the oxygen level is cut in half, the hemoglobin carries only about one-quarter the amount of oxygen. This makes it easier for hemoglobin to pick up oxygen in the lungs and release it in the tissues. The presence of carbon dioxide also affects the ability of hemoglobin to carry oxygen. At higher carbon dioxide levels, the hemoglobin molecule is distorted in shape and doesn't bind to oxygen as well (the Bohr effect). This also helps oxygen leave the blood in the tissues.

    (ii) Most carbon dioxide is found in the cytoplasm of the RBC. Carbonic anhydrase, an enzyme present in the RBC, causes carbon dioxide to form carbonic acid with water. This acid breaks down into protons (like other acids) and bicarbonate ions (HCO_3^-). As the carbon dioxide is changed into bicarbonate ions, the level of carbon dioxide in the blood falls and more carbon dioxide diffuses out from the tissues. Since there is only a small (about 5 percent) difference in the concentration of carbon dioxide between the blood and the tissues, this helps the blood carry more carbon dioxide. About 8 percent of the carbon dioxide carried in the blood is dissolved in the plasma, and the remaining 20 percent is bound to hemoglobin. At the lungs, the same carbonic anhydrase reaction causes bicarbonate to release carbon dioxide. The carbon dioxide carried by the hemoglobin is released since hemoglobin has a greater attraction (affinity) for oxygen than for carbon dioxide. **(Respiration, gas exchange, control)**

40. A mature red blood cell has no mitochondria, but a great deal of hemoglobin. Hemoglobin is a large, complex iron-containing molecule consisting of four polypeptide subunits. RBCs produce ATP by anaerobic means. Mature mammalian RBCs are unique among the vertebrates in that they have no nuclei. Under the microscope, RBCs look like donut or bagels with a depression rather than hole in the center. In cross section, they look something like a dog's bone. **(Blood)**

41. (i) A fish gains oxygen from the water and loses carbon dioxide through gills, which are folded, blood-rich filaments over which water is forced. The continuous flow of fresh water through the gills helps gas exchange. Blood flows in a direction opposite to the flow of the water, so that the blood with the least oxygen meets water with the least oxygen, and blood with a higher concentration of oxygen meets water with more oxygen. In this way, the difference in oxygen concentration (a gradient) between the blood and the water is maintained at a maximum, favoring diffusion of oxygen into the blood at all points in the gills. This is called countercurrent flow, and makes fish gills very efficient respiratory organs.

    (ii) An adult frog has simple lungs made up of folded membrane, which increases the surface area available for gas exchange. The lungs are assisted by gas exchange through the moist skin. The skin helps whether the frog is in the air or the water. Frogs force air into their lungs by swallowing, an action that is readily apparent if a frog is closely examined.

(iii) Air enters through the nose, moves down the trachea and into the various sections of the lungs and finally into little elastic sacs called alveoli. There gas exchange occurs. The alveoli greatly increase the surface area of the lung; an adult human has something like the area of football field available for gas exchange. During inhalation, the rib cage moves up and outward due to the contraction of muscles, the diaphragm (a muscular, dome-shaped sheet lying over the liver and partitioning the thoracic and abdominal cavities) contracts and moves down. As a result there is more space in the chest, the lungs are at a lower pressure than atmospheric pressure, and air rushes in to the lower-pressure region. Exhalation forces air out by increasing the air pressure in the lungs to above atmospheric pressure; the diaphragm relaxes and the rib and abdominal muscles contract.

(iv) Birds need a lot of energy in order to fly and maintain their body temperature. Thus they need efficient respiration. When a bird inspires, the air moves into posterior air sacs. Exhaling, the air moves into the lungs. On a second inspiration, air moves out of the lungs into anterior air sacs, where it is expelled from the body on the second exhalation. By this process, birds have a one-way flow of air. The air is almost completely changed as the bird breathes. In mammalian lungs, by contrast, there is dead space. In mammals, the air is never completely changed in breathing as the lungs and trachea always retain some old air. It is important to note that the air sacs function as bellows only—they are not used in gas exchange. Their function is to provide the lungs with fresh air. **(Respiration, gas exchange, control)**

42. (i) False

    (ii) False

    (iii) True

    (iv) True **(Respiration, gas exchange, control)**

43. b and d are true **(Respiration, gas exchange, control)**

44. alveoli **(Respiration, gas exchange, control)**

45. b. diffusion. Gas exchange, in the end, is all by diffusion. Lungs and gills provide a means of bringing blood and air or water together to effect an exchange, but the final step is always exchanged by diffusion. Some mechanisms, such as the counter-current exchange arrangement found in gills, or air sacs in birds, have evolved to improve the efficiency of the system. **(Respiration, gas exchange, control)**

46. breathing center in the medulla. Respiration is controlled by monitoring the concentration of oxygen, carbon dioxide, and acid (pH) in the blood. Control is located in the brain, and in mammals, in receptors in the carotid arteries of the neck. **(Respiration, gas exchange, control)**

47. Spiders breathe with what are called book (or sometimes diffusion) lungs. **(Respiration, gas exchange, control)**

48. The partial pressure of a gas is the portion of a total pressure of a gas mixture that is contributed by a specific gas in that mixture. For instance, there is about 21 percent oxygen in the air, and 21 percent of the total air pressure (one atmosphere or 760 Torr) is due to oxygen. Partial pressure is a way of talking about the concentration of a gas. The higher the partial pressure, the more gas there is of that type. **(Respiration, gas exchange, control)**

49. b. fibrinogen is converted to fibrin in a series of reactions and the fibrin holds back cells. **(Blood)**

50. Blood clots by a complex mechanism, which may be summarized as follows. Circulating platelets adhere to the damaged wall of a blood vessel which has been severed. These platelets release a factor which makes nearby platelets sticky, causing the platelets to clump together and seal the blood vessel. The

platelets also release clotting factors which, along with substances such as vitamin K (present in the plasma), cause protein, prothrombin, to become the active form, thrombin. Thrombin in turn converts fibrinogen to fibrin. Fibrin threads or fibers form a sieve to trap cells and stop bleeding. At this point, healing starts. **(Blood)**

51. In a person, during inhalation, the rib cage move up and outward due to the contraction of muscles, the diaphragm contracts and moves down. As a result there is more space in the chest, the lungs are at a lower pressure than the atmosphere, and air rushes in. Exhalation forces air out by increasing the air pressure in the lungs to above atmospheric pressure.

    A frog swallows air to expand the lungs. **(Respiration, gas exchange, control)**

52. Chemoreceptors in the carotid (neck) arteries monitor the level of the blood oxygen. At the same time, chemoreceptors in the brain and in the carotid arteries detect the level of carbon dioxide in the blood as well the pH (pH falls as carbon dioxide level rise—more protons are lost from carbonic acid). This is why a person who is hyperventilating from anxiety should breath into a paper bag. The too-rapid breathing blows off (releases) too much carbon dioxide from the blood, and as the blood pH rises, the person feels faint. This often provokes them to breathe even faster. In the bag, carbon dioxide builds up, the level of carbon dioxide in the blood increase, the blood pH drops, and the dizzy sensation passes. **(Respiration, gas exchange, control)**

53. All gas exchange must occur through a surface. (This can be imagined as a window through which oxygen and carbon dioxide move.) The amount of gas that can be exchanged is dependent on the surface area (size of the window). The area of a surface can be increased by folding or forming pockets. As an example, you can accordion fold a legal-sized sheet of paper to occupy the same space on a desk as a regular sheet of notebook paper. The pages occupy the same desk space but the surface area (amount of paper) is greater for the folded sheet. **(Respiration, gas exchange, control)**

54. Blood and the circulatory system transport gases, nutrients, ions, hormones and water throughout the body. The circulatory system also distributes body heat. This all contributes to maintaining the body's internal environment. **(Homeostasis)**

55. a. eating a balanced diet. Homeostasis is the maintenance of the internal equilibrium of an organism by keeping pH, oxygen and carbon dioxide levels, ion concentrations, blood glucose levels, within limits necessary for normal physiology. **(Homeostasis)**

56. Respiration as described in the answer to question #52 above, is a good example of corrective action taken to keep the body homeostatic. **(Homeostasis)**

57. The concept of homeostasis is due to (d) Claude Bernard; (b) Walter B. Cannon coined the term. **(Homeostasis)**

58. Blood sugar levels are maintained within a certain range. If the level drops too far, faintness is felt. This is called hypoglycemia. The body reacts by releasing glucose from the liver. The pH of the blood is also maintained, both through respiratory and metabolic controls. A third example would be the balance between water and salt. When the water levels in the body drop, we feel thirsty. Drinking brings more water into the system. When salt concentrations drop, loss of salt through the urine is reduced. Homeostasis is a lot like maintaining a constant balance in a checking account. As money is spent, more money needs to be added to maintain a constant balance. **(Homeostasis)**

59. Lymph is a fluid similar to plasma in that it is cell-free and contains many substances such as proteins. Fluid that leaves blood capillaries and enters the tissues is returned to the circulation as lymph. **(Lymphatic system)**

60. The fluid first drains into lymph passages (lymph capillaries) and flows through lymph nodes (where white cells monitor it for signs of infection and remove debris), ending up in the large thoracic duct where the lymph drains into the subclavian veins that lie below the collar bone (clavicle). Fluid leaving the blood capillaries finds its way back from the tissues as lymph. The lymphatic system maintains electrolyte and fluid balance in the body, and assists in the work of the immune system. It drains tissue spaces and cavities of fluids not recovered by the capillaries. Lymph is moved mainly by muscle action and pressure changes within the thorax. **(Lymphatic system)**

Grade Yourself

Circle the question numbers that you had incorrect. Then indicate the number of questions you missed. If you answered more than three questions incorrectly, you will then have to focus on that topic. (If a topic has less than three questions and you had at least one wrong, we suggest you study that topic also. Read your textbook, a review book, or ask your teacher for help.)

Subject: Homeostasis, Respiration, and Circulation

Topic	Question Numbers	Number Incorrect
Body systems	1	
Circulatory systems	2, 6, 7, 8, 9, 12, 13, 22, 28, 29, 30, 33, 34, 37, 38	
The heart	3, 4, 5, 10, 11, 15, 16, 17, 18, 23, 24, 27, 32, 36	
Blood	14, 19, 20, 21, 31, 35, 40, 49, 50	
Control of circulation	25, 26, 28	
Respiration, gas exchange, control	35, 39, 41, 42, 43, 44, 45, 46, 47, 48, 51, 52, 53	
Homeostasis	54, 55, 56, 57, 58	
Lymphatic system	59, 60	

The Diversity of Plants: Classification, Life Cycles, and Reproduction

Brief Yourself

Plants—General

Plants are eukaryotic organisms, generally multicellular, sessile, with cellulose cell walls and cell plastids. They usually produce starches as their storage carbohydrate. Most plants are photosynthetic autotrophs, possessing chlorophylls *a* and *b* as well as accessory pigments such as carotenoids. Many plants exhibit alternation of generations in which they switch between haploid (1 N) and diploid (2 N) forms during their life cycle.

Modern plants are thought to have evolved some 500 million years ago from green algae which were the ancestors of the modern denizens of Division Chlorophyta. While adapting to a terrestrial habitat, plants faced a number of new challenges; land plants, for instance, must resist drying, require a means of transporting gametes so as to ensure fertilization, and need to move food and water throughout the plant body. In the course of addressing these problems many of the characteristics of modern plants were established.

Plants—Taxonomy

Plants are usually divided into two main groups, the Bryophytes and the Tracheophytes. Note that *class* and *division* are equivalent taxonomic terms used by different sources (e.g., Class Musci or Division Musci both refer to mosses).

I. Bryophytes (lack well-developed vascular tissue)
   - A. Mosses (Division Musci)
   - B. Liverworts (Division Hepaticae)
   - C. Hornworts (Division Anthocerotae)

II. Tracheophytes (have vascular tissue)
   - A. Without seeds (sometimes termed ferns and fern allies)
     1. whisk ferns (Division Psilophyta)
     2. club mosses (Division Lycophyta)

3. horsetails (Division Sphenophyta)
4. ferns (Division Pterophyta)

B. With seeds

1. conifers (Division Coniferophyta)
2. cycads (Division Cycadophyta)
3. ginkos (Division Ginkgophyta)
4. gnetophytes (Division Gnetophyta)
5. angiosperms (Division Anthophyta)
   a. dicots (Division Dicotyledonae)
   b. monocots (Division Monocotyledonae)

Test Yourself

1. List three features that distinguish plant from animal cells.

2. Name one difference between fungal and plant cells that does not relate to cell organelles.

3. Give a common group name (and if possible, an example) for each of these divisions (e.g. "flowering plants" and "oak tree" for Angiosperms). For each of the starred (*) items, write a brief description of that group.
   a. Bryophyta*
   b. Lycophyta*
   c. Sphenophyta*
   d. Psilophyta*
   e. Pterophyta*
   f. Cycadophyta
   g. Ginkgophyta
   h. Coniferophyta*
   i. Gnetophyta
   j. Anthophyta *

4. Plants which live on other plants without being parasites are called ____________.

5. The earliest seed plants were the ____________.

6. ______________ are horizontal underground stems that function like roots.

7. Which has a larger gameteophyte relative to its sporophyte, a fern or an angiosperm?

8. Nonvascular plants in existence today include only the ______________.

9. The microgametophyte of angiosperms is ____________.

10. When moss spores germinate they form ___________.

11. How does sperm motility differ between ferns and conifers or angiosperms?

12. _______________ (common name) contain silica and were used at one time like sandpaper.

13. Mosses produce only ____________ which can give rise to either female or male gametophytes. Vascular plants, on the other hand, produce both _______________ and ______________ which respectively produce male and female gametophytes.

14. The green moss on a rock is a ____________ with haploid cells, while the daisy is a __________ with diploid cells.

15. The pollen producing cones in conifers are __________ (smaller/larger) than the ovule-bearing cones.

16. What tissue is triploid in angiosperms? Why doesn't this triploidy cause genetic problems in subsequent generations?

17. Terrestrial plants evolved from what group of plants (or protists)?

18. ________ spore (hetero-, homo-) plants produce two distinct spore types, ______________ which germinate into male gametophytes, and ____________ that germinate into female gametophytes. Which spores are smaller?

19. Conducting tissues are found in vascular plants collectively known as ____________.
    a. psilophyta
    b. bryophyta
    c. chordata
    d. tracheophyta
    e. angiospermophyta

20. Conducting tubes which transport carbohydrate food are called ___________; the transport vessels for water are ____________.

21. Which would be harder to distinguish by eye, the gametophyte and sporophyte forms of a isomorphic or heteromorphic plant?

Questions 22–30. True or false? If false, correct to make the statement true.

22. Haploid spores combine to form diploid cells which divide by mitosis.

23. A diploid adult is a sporophyte plant.

24. A haploid adult is a sporophyte plant.

25. Gametophytes make gametes by mitosis.

26. Sporophytes make spores by mitosis.

27. Alternation of generations involves only sexual reproduction.

28. Gymnosperms do not have motile sperm.

29. Mosses attach to rocks, soil, and other substrate with mycorrhizae.

30. All plant cells possess chloroplasts.

31. Modern plants are thought to have evolved some _____ million years ago from green algae .
    a. 200
    b. 300
    c. 400
    d. 500

32. True or false—If false correct to make the statement true. In mosses, antheridia produce eggs, while archegonia produce sperm.

33. Ferns have leaves called _________.

34. On the underside of the leaves of ferns are small dot-like structures called __________. These are made up of sporangia which burst open releasing ___________ (haploid/diploid) spores.

35. The structure which produces spores in the plant *Lycopodium* is the __________.

36. The gametophyte generation is most reduced in the ____________.
    a. bryophytes
    b. ferns and allies
    c. gymnosperms
    d. angiosperms

37. Seed-bearing plants include (list all that apply):
    a. bryophytes
    b. ferns and allies
    c. gymnosperms
    d. angiosperms

38. Plants that have vascular tissue but do not produce seeds are (list all that apply):
    a. bryophytes
    b. ferns and allies

c. gymnosperms

d. angiosperms

39. Which plants produce fruit? (List all that apply.)

    a. bryophytes

    b. ferns and allies

    c. gymnosperms

    d. angiosperms

40. Most plants species now in existence are:

    a. bryophytes

    b. ferns and allies

    c. gymnosperms

    d. angiosperms

41. Moving from the outside in, place the following flower structures in the correct order: petals, stamens, carpels, sepals.

42. ________ produce pollen, while _________ are the female structures.

43. Flowers which possess __________ and __________ in the same flower are termed perfect.

44. The ovule develops into the _________, and the surrounding _________ into the fruit.

45. In the stamen, pollen is formed in the __________ which sits on top of a stalk, the ____________.

46. Plants like corn which have separate male and female flowers on the same individual plant are designated ___________. If the male and female flowers occur on different individuals, they are considered __________.

47. Which of the following describes

    a. typical monocots _________

    b. typical dicots _________

    (i) Leaves with branching or netlike patterns of veins, flower parts occurring in fours or fives (or multiples thereof), broad leaves, vascular bundles in the form of rings.

    (ii) Broad leaves with parallel veins, flower parts in fours or fives (or multiples thereof), fibrous roots, and scattered vascular bundles.

    (iii) Narrow leaves with parallel veins, flower parts in threes (or multiples thereof), fibrous roots, and scattered vascular bundles.

48. Lilies, grasses and grains are examples of __________. Woody trees, shrubs and plants with taproots (like carrots) are __________.

49. Seed-bearing plants first appeared about _________ million years ago.

    a. 100

    b. 175

    c. 225

    d. 360

50. Write a short essay discussing the major features of alternation of generations in plant life cycles.

51. (i) What plant features (such as roots) are necessary or helpful in allowing plants to colonize the land?

    (ii) Discuss seeds as an important development in the evolution of plants.

52. Discuss the general features of the seedless, vascular plants.

53. Which plants are heterosporous?

54. Briefly discuss the advantages seed-producing plants have over seedless plants.

55. A complete flower has all four parts, _______ , _______ , _______ and _______ . A perfect flower is bisexual, meaning it has _______ and _______ . (Possible terms: pistil, stamen, petal, sepal, carpel)

Check Yourself

1. Plant cells have a cell wall, possess chloroplasts or other plastids, and lack centrioles (found in animals cells but not plant cells during cellular division). **(Plants—general characteristics and origin)**

2. fungal cells have cells wall of chitin, not cellulose. **(Plants—general characteristics and origin)**

3. a. Bryophyta. **(Bryophytes)**

   Plants of this division (over 16,000 species) lack vascular tissues, true roots, or cuticles impermeable to water; water reaches tissues through diffusion and capillary action (like a paper towel wicks up water), and is circulated through cells by cytoplasmic streaming. Bryophytes are of necessity mostly small plants, and actively grow and reproduce only in wet environments or during wet periods. Their sperm must swim to reach the eggs, and thus bryophytes are not totally adapted to independent life on land.

   Mosses (class Musci, over 10,000 species) are abundant in all climates, even the arctic, but are dormant during dry or very cold periods. Mosses attach to rocks, tree trunks, and other surfaces with root-like structures called rhizopods which also serve to absorb water. The haploid stage is the dominant and most visible stage in the moss life cycle. When you look at moss, you are viewing a gametophyte composed of haploid cells. At the tips of shoots, gamatangia called antheridia (singular: antheridium) produce sperm by meiosis, while those called archegonia (singular: archegonium) produce eggs. Wet conditions are necessary for the sperm to swim to the egg. The resulting haploid spores are scattered by the wind. Those that land in favorable environments become protonemata, which often looks like an area of green discoloration due to algal growth. Protonemata produce the familiar male and female gametophytic generation.

   The less common bryophytes include the liverworts (class Hepaticae, with 6,500 species), with lobed green plant bodies, and hornworts (class Anthocerotae with 100 species) which have cells possessing a single large chloroplast.

   b. Lycophyta (clubmosses, *Lycopodium*), c. Sphenophyta (horsetails).

   The horsetails (Sphenophyta) and club mosses (*Lycophyta*) have an ancestral line reaching back hundreds of millions of years. These are the first plants with true roots and structures akin to leaves. During the Carboniferous period (350- 280 million years ago) these plants flourished. Some clubmosses were the size of present day trees. Rhizomes of sporophyte clubmosses have roots and stems with vascular tissue. A common form, *Lycopodium*, like many other lycopods, has a special compact leaf structure for the production and dispersal of spores, called a strobilus (plural strobili). Some lycopods are adapted to living on other plants for support but not for nutrition (i.e. they are not parasites). These type of plants are called epiphytes. Club mosses are represented by five genera today, and less than 1000 species. Horsetails are represented today by a single surviving genus, *Equisetum*. Vegetative stems are green, often with a feathery appearance; reproductive stems bear the aforementioned strobili and are non-photosynthetic and pale. Horsetails have silica particle in their stems, an adaptation which discourages herbivores and allowed colonial American woodworkers to employ them as sandpaper, and by otheres as pot cleaners.

   c. Psilophyta (whiskferns).

   Whiskferns (Psilophyta) look like whiskbrooms, and are the least advanced of the group. Whiskferns have cells with unusually large chromosomes, rhizomes (horizontal, underground stems) which function like roots and have asymbiotic relationship with mycorrhizial fungi. They probably are direct descendants of the earliest land plants, and as a group date back perhaps 400 million years.

   d. Pterophyta (ferns, Boston fern).

Ferns (Pterophyta) are the most familiar and prevalent (about 11,000 species) of the modern seedless vascular plants. Ferns range in size from under a centimeter to more than 25 meters tall. The size of the latter is of course possible due to the presence of vascular transport tissues. Most ferns, mainly excepting the epiphytic forms, have below-ground rhizomes from which arise leaves (fronds) and roots. Pinhead-like sori (singular sorus) can be seen on the underside of fronds. Sori consist of sporangia (singular sporangium) which burst open to disperse the haploid spores which have been developing inside. Each spore which germinates gives rise to a haploid gametophyte plant. Flagellated sperm form in the antheridium and swim in water to the egg, which lies sessile in the archegonium. After the egg and sperm unite as a zygote, the diploid sporophyte develops while still attached to the gametophyte.

e. Cycadophyta (cycads, sago palm), f. Ginkgophyta (ginko), g. Coniferophyta (conifers, pines), h. Gnetophyta (Mormon tea plant.)

The earliest seed-bearing vascular plants were gymnosperms, which include conifers (pines, hemlocks, spruces, redwoods, etc.). Sporophyte-stage trees produce cones, reproductive organs derived from leaves. Many species have two types of cones, the smaller male or pollen cones which produce haploid microspores which become the pollen. Haploid megaspores are produced by the larger female or ovulate cones, and develop into female gametophytes. (A useful generalization: in reproductive biology, the prefix *micro* is applied to the smaller male cells, while *mega* is used to refer to larger female cells.) Male gametes (pollen) are blown off of the male cones, and pollinate the ovules located on the scales on the ovulate cones. As the pollen grains germinate, they grow long pollen tubes into the female tissues, where the sperm fertilize the female gametophytes.

The sporophyte stage is the most apparent, with the haploid, gametophyte stage being limited to male microspores and female megaspores (and the resulting female gametophyte). Once fertilized, the diploid sporophyte phase starts again in the form of a zygote.

Some gymnosperm sperm are flagellated, but are carried to near the egg cell by the growing pollen tube, rather than by swimming as in lower plants. Conifers and Gnetophyta have non-motile sperm.

i. Anthophyta (flowering plants, daisy)

Perhaps the most familiar of plants, flowering plants have come to dominate the landscape in the past 150 million years—indeed, most modern plants are angiosperms. In Angiospermophyta, the gametophyte generation is confined to the reproductive structures. The plant bodies we are familiar with are the dominant sporophytes. Flowers are one of the great success stores of the plant world. In place of large quantities of pollen distributed by wind, as is the common mode of pollination in the gymnosperms, or the dependence of lower plants on water for the transmission of gametes, angiosperms have coevolved a close relationship with insects, birds, and bats. These animals serve as the vehicles of pollination.

The diversity of flowers is amazing, but a generalized flower may be usefully described. A flower consists of a number of parts. Complete flowers are those defined as having sepals (green, leaf-like structures at the base of the flower, which protected the bud before it opened), petals, stamens (male structures made up of the stalk-like filament and the anther where pollen is formed), and carpels (female structures). Any flower which contains both stamens and carpels is called perfect. The single carpel or group of fused carpels has an uppermost stigma, where pollen grains are transferred (perhaps from an insects body) to the tacky surface. The pollen grain germinates there, and grows a pollen tube down the style (the filament supporting the stigma). Here double fertilization occurs. The nucleus of the egg cell is fertilized by one sperm. Another sperm fuses with the two polar nuclei to form the triploid (3N) endosperm; the endosperm grows and is later used by the embryo as food. The ovule develops into the seed, and the surrounding ovary into the fruit. In many angiosperms, the fruit is a conspicuous attractant for animals who can spread seeds collected on their fur or in their digestive tracts. Plants like corn which have separate male and female flowers on the same individual plant are designated monoecious. If the male and female flowers occur on different individuals, they are considered dioecious **(Plant taxonomy)**

4. epiphytes **(Plants—general characteristics and origin)**

5. gymnosperms **(Gymnosperms)**

6. Rhizomes **(Ferns and fern allies)**

7. A fern. Although the gametophyte in ferns is much smaller than the sporophyte, the gametophyte in angiosperms is limited to a group of a few cells. Only in bryophytes the sporophyte is usually smaller than the gametophyte stage. **(Ferns and fern allies)**

8. brophytes **(Bryophytes)**

9. pollen **(Angiosperms)**

10. protonema **(Bryophytes)**

11. Fern sperm bear flagella and are therefore motile; the pollen of conifers and angiosperms is nonmotile and must be passively borne to a place near the egg cell. **(Ferns and fern allies)**

12. Horsetails.**(Ferns and fern allies)**

13. homospores or microspores, microspores, megaspores **(Bryophytes)**

14. gametophyte, sporophyte **(Plant life cycles)**

15. smaller **(Gymnosperms)**

16. The endosperm tissue in the seed results from the fusion of three haploid cells, and is triploid. Its triploidy is not inherited as the endosperm is consumed by the embryo and contributes no genetic material to future generations. **(Angiosperms)**

17. green algae **(Plants—general characteristics and origin)**

18. heterospore, microspores, megaspores. As the names suggest, microspores are the smaller of the two. **(Plant life cycles)**

19. d. tracheophyta **(Developments in evolution)**

20. phloem, xylem. An old memory trick is to associate by sound the word "***ph***loem" with "***f***ood." **(Developments in evolution)**

21. The isomorphic plant—it would be the hardest to identify the different forms in. Recall *iso-* refers to same, while *hetero-* refers to different. **(Plant life cycles)**

22. False. Haploid spores do not combine to form diploid cells; they do divide mitotically. **(Plant life cycles)**

23. True **(Plant life cycles)**

24. False A haploid adult is a gametophyte plant. **(Plant life cycles)**

25. True **(Plant life cycles)**

26. False. Recall that it is diploid sporophyte adults which produce haploid spores by meiosis. **(Plant life cycles)**

27. False. In alternation of generations, both sexual and asexual modes of reproduction are used. At some point sexual reproduction is necessary to reconstitute the diploid genotype if meiosis is to occur. **(Plant life cycles)**

28. False. Gymnosperms such as the cycads and the ginkos have motile sperm. The sperm are carried to near the egg by the pollen tube, rather than by the sperm's direct efforts. **(Gymnosperms)**

29. False. Mosses attach to rocks, soil, and other substrates with rhizopods. Mycorrhizae are fungi that live in symbiosis with the rhizomes of plants like whiskbrooms. **(Bryophytes)**

30. False. All plant cells possess plastids, but not necessarily in the form of chloroplasts. **(Plants—general characteristics and origin)**

31. d. 500 **(Plants—general characteristics and origin)**

32. False. In mosses, antheridia produce sperm and archegonia produce eggs. **(Bryophytes)**

33. fronds. **(Ferns and fern allies)**

34. sori, haploid. **(Ferns and fern allies)**

35. strobilus. **(Ferns and fern allies)**

36. d. angiosperms **(Plant life cycles)**

37. c. gymnosperms and d. angiosperms **(Gymnosperms)**

38. b. ferns and allies **(Ferns and fern allies)**

39. d. angiosperms **(Angiosperms)**

40. d. angiosperms **(Angiosperms)**

41. sepals, petals, stamens, carpels **(Angiosperms)**

42. Stamens, pistils **(Angiosperms)**

43. stamens, pistils **(Angiosperms)**

44. seed, ovary **(Angiosperms)**

45. anther, filament **(Angiosperms)**

46. monoecious, dioecious **(Angiosperms)**

47. a. iii, b. i. To help keep the monocots and dicots straight, remember that monocots have the least number of cotyledons (one) and the smallest number (three) associated with their flowers; typical monocots include lilies, grasses, and grains. Dicots, as the name suggests, have the larger number of cotyledons

(two) and the larger number of flower parts (four or five), and are represented by the woody trees and shrubs. **(Angiosperms)**

48. monocots, dicots. (1) The monocotyledons ("monocots") produce embryos with a single cotyledon or seed leaf. Monocot plants typically have narrow leaves with parallel veins, flower parts that occur in threes or multiples of three, a fibrous root system, and scattered vascular bundles that form complex patterns. Examples of monocots include lilies, grasses and grains, orchids and less commonly trees like palms. Today there are more than 65,000 species of monocots. (2) The dicotyledons ("dicots") produce embryos with two cotyledons. Dicots typically have broad leaves and branching or netlike patterns of veins. Flower parts occur in fours or fives, or multiples of these. Taproots (e.g., carrots) are common, and vascular bundles usually form a ring. Cambium is frequently present in dicots, but rare in monocots; dicots include the woody trees and shrubs. There are estimated to be about 160,000 dicot species extant. **(Angiosperms)**

49. c. 225 **(Angiosperms)**

50. The biologist Lynn Margulis has remarked that since some plants are devoid of photosynthetic activity, what truly distinguishes plants from members of other kingdoms is their life cycles. Certainly plants are characterized by unique life cycles, as well as a universally sessile nature (animals move to get places; plants grow). Plants have a life cycle of alternating haploid (gametophyte) and diploid (sporophyte) generations. The dominant, or highly visible form varies. The gametophyte generation is dominant in the older groups of plants such as the bryophytes, whereas the other groups that have appeared more recently are predominately sporophytes in their life cycles; thus the green moss is a gametophyte but an oak tree is a sporophyte. The bryophytes now remain as the one major exception to the evolutionary tend toward emphasizing the sporophyte generation. Alternation of generations involves sexual and asexual reproduction.

    In some organisms like the alga *Ulva* (which may be classified as protists) the adult's gametophyte and sporophyte forms are visually indistinguishable. This identity of form is called isomorphy (adjective—isomorphic). Most plants, by contrast, have readily distinguished gametophyte and sporophyte stages—these plants are heteromorphic.

    The gametophyte plant produces gametes by mitosis; meiosis is unnecessary since the cells are already haploid. The sporophyte produces spores by meiosis since sporophyte cells are diploid. The transition from gametophyte to sporophyte starts with fertilization of the egg by sperm (syngamy), whereas the gametophyte generation starts when the haploid spores are produced. **(Plant life cycles)**

51. There are a number of developments in the evolution of plants that have been important in establishing modern plants around the world.

    (i) Terrestrial colonization has been dependent upon such things as adaptations as roots (which anchor and allow the plant to efficiently take up nutrients and water) water conservation features (leaf cuticle and stomates that allow control of water loss during gas exchange), support tissues (which make larger plants possible, e.g., secondary growth tissue), transport tissues (another important factor in plant size), specialized organs such as leaves, and means of sperm dispersal independent of a film of water. Bryophytes and lower tracheophytes such as ferns show fewer of these adaptations than do gymnosperms and angiosperms.

    (ii) Seeds merit special mention. Compact, resistant to drought and idle predation, packed with nutrients and energy to feed the embryo's development into a seedling, seeds are well-suited to fostering the wide dispersal of plants, and along with flowers, do much to account for the present day predominance of angiosperms. In the seed, the gametophyte forms consist of a few cells, surrounded by and dependent upon sporophyte tissues. **(Plants—general characteristics and origin)**

52. Seedless vascular plants: Ferns and fern allies. These plants are the first example of the tracheophytes, plants which use vascular tissues to transport water and food throughout the plant body. Transport (conductive) tissues are of two types: (1) tracheids (or tracheary or vessel elements: tube-shaped elements with hard, generally waterproof walls, e.g., xylem for conducting water) and (2) sieve tube elements (also tubular, with soft walls and perforations, e.g., phloem for conducting carbohydrate food). **(Plants—general characteristics and origin)**

53. Only vascular plants are heterosporous (as opposed to homosporous)—i.e., have spores that come in two different sizes. These are the larger megaspores that become female gametophytes, and smaller microspores become the male gametophytes. **(Plants—general characteristics and origin)**

54. Seed plants have evolved several major advantages over the seedless plants, and these adaptations have served the seed-bearing plants well over the past 360 million years. First, sperm are carried as pollen by wind, birds or insects, freeing these plants from the need for water in order to accomplish fertilization. This is a major advantage in a terrestrial habitat. Ovules in seed bearing plants encompasses a female gametophyte which produces the egg. When fertilized, the eggs are encapsulated within a seed coat, a sealed packet containing the embryo and the nutrients which will feed its initial development. This is a second major advantage as the hardy, dormant seeds are also well-suited to a terrestrial environment. **(Plants—general characteristics and origin)**

55. pistil, stamen, petal, sepal, pistil, stamen **(Angiosperms)**

Grade Yourself

Circle the question numbers that you had incorrect. Then indicate the number of questions you missed. If you answered more than three questions incorrectly, you will then have to focus on that topic. (If a topic has less than three questions and you had at least one wrong, we suggest you study that topic also. Read your textbook, a review book, or ask your teacher for help.)

Subject: The Diversity of Plants: Classification, Life Cycles, and Reproduction

Topic	Question Numbers	Number Incorrect
Plants—general characteristics and origin	1, 2, 4, 17, 30, 31, 51, 52, 53, 54	
Plant taxonomy	3	
Gymnosperms	5, 15, 28, 37	
Ferns and fern allies	6, 7, 11, 12, 33, 34, 35, 38	
Bryophytes	8, 10, 13, 29, 32	
Angiosperms	9, 16, 39, 40, 41, 42, 43, 44, 45, 46, 47, 48, 49, 55	
Plant life cycles	14, 18, 21, 22, 23, 24, 25, 26, 27, 36, 50	
Developments in evolution	19, 20	

Plants: Structure, Growth, Development, and Physiology

Brief Yourself

Vascular Plant Anatomy

A plant is anchored to the ground by its roots, and uses these to obtain nutrients and water. The stem and branches, forms of shoots, reach up, and lastly become leaves and in angiosperms, flowers. Vascular plants have a special region at the tips of roots and stems called the apical meristem which is responsible for changes in length. Another region called the lateral meristem, produces enlargement in girth.

Types of Plant Cells and Tissues

Primary meristem, present in the angiosperm embryo, gives rise to three types of tissue: dermal (outer covering), vascular (transport tissues, xylem and phloem), and ground tissue (the matrix in which vascular tissues exist). The term meristem refers to cells that are in an undifferentiated state, sort of "all-purpose cells." Higher plants have differentiated tissues in the form of organs (leaves, shoots, roots, and in angiosperms, flowers) each with specialized functions.

Sclerenchyma cells with secondary cell walls reinforced by lignin, which lends rigidity to the plant, are found in the ground tissue of stems. At maturity, these cells are dead. Collenchyma cells are often found in groups under the epidermis in stems; the cell walls are unevenly thickened and give additional strength to stems and petioles. Collenchyma cells lack secondary walls and are important in the growth of young plants. Not as rigid as schlerenchyma, at maturity they are alive and functioning. Parenchyma cells are unspecialized meristematic cells, and can differentiate into other types of cells. Epidermal cells cover the exterior of the plant, and endodermis cells are common in roots.

Growth

The meristematic regions in the shoot and root tips are responsible for all primary growth. In primary growth, the plant body elongates and meristem cells undergo differentiation into various tissues. Such tissues come from the differentiation of meristem into primary meristems. These are protoderm (becomes epidermis), ground meristem (becomes pith and cortex tissues), and procambium (becomes xylem and phloem). New leaves come from the apical meristem in the shoot, specifically in the leaf primordia. Stems grow through terminal buds which produce nodes and internodes during the spring and summer growth. Axillary buds occur above the points on the stem where leaves were formerly attached. The apical buds hormonally inhibit the growth of axillary buds, an action referred to as apical dominance. Growth in plants is functionally analogous to the behavior of animals; as animals move, plants grow to change their relationship to their surroundings.

Plant Hormones

Plant hormones are chemical messengers which in minute quantities coordinate activities in different parts of the plant and affect how and when cells divide, elongate and differentiate. Charles Darwin and his son, Francis, were the first to propose the existence of a plant hormone (auxin) which caused certain areas of the stem beneath the plant tip to elongate. Shielding tips of canary grass seedlings with little caps of lead foil, and exposing the plants to light on only one side, they showed that the seedlings bend toward light only when the coleoptile (which forms around the epicotyl in grass seeds) uncovered.

Water and Nutrients

Plants must absorb water and mineral elements from the soil. Soil varies tremendously in its makeup and resulting character; clay particles hold onto water through hydrogen bonding, while sand freely passes water. The composition of soil in terms of particle type and size will determine its ability to hold water (its field capacity).

Test Yourself

1. What are three functions of roots?

2. A plant is placed next to a window, and the stem bends toward the light. This response is most likely due to:

   a. cytokinins

   b. auxin distribution

   c. ethylene concentration

   d. temperature

3. List four plant macronutrients.

4. List four plant micronutrients.

5. Which two of the following plant hormones have an antagonistic relationship in determining plant responses?

   a. ethylene

   b. cytokinins

   c. auxins

   d. abscisic acid

6. What ion is actively pumped into guard cells? What ion can passively follow and prevent the establishment of a electrochemical gradient across the membrane?

7. How is most water lost by plants?

8. What is the most common form of carbohydrate transported with plants?

9. Give one reason why soil pH affects plant growth.

10. Carbohydrate is made at _________ and used or stored at _________.

11. What is field capacity? What soil factors affect field capacity?

12. Bacteria of what genus fix atmospheric nitrogen?

13. Plants are __________ phototrophic but negatively _____________.

14. Which hormones induce leaves to drop from plants?

15. Which lies closer to the center of a tree, sapwood or heartwood?

16. Flowers are evolutionarily derived from what plant structures?

17. List these monocot root structures in proper order, working from the outside of the plant to the inside.

    root hairs, pith, pericycle, epidermis, vascular cylinder, cortex

18. Describe the process of germination, and the three parts of the seed embryo. What are two unusual conditions some seeds require in order to germinate?

19. Secondary growth refers to growth that increases the _________ of a tree.

20. What causes apical dominance?

For questions 21–26, answer with the correct cell type. Choices are: sclerenchyma, collenchyma, parenchyma, epidermis, endodermis.

21. Forms fibers used in rope. Mature cells not living.

22. Remain undifferentiated and can thus become other types of cells.

23. What distinguishes a pear by its "grittiness" when chewed.

24. Root hairs are part of these.

25. Contains lignin in the secondary cell wall.

26. Found in roots, where suberin waterproofs them so as to form a gasket between the exterior and interior tissues of the root.

27. True or false: Xylem is made up of dead cells.

28. Tracheids are the only conducting cells usually found in (list all that apply):

    a. gymnosperms

    b. angiosperms

    c. ferns

    d. bryophytes

29. The gasket that encloses the center of the root and forces water and solutes to take a route across cell membranes is called the:

    a. stele

    b. vascular cylinder

    c. Casparian strip

    d. spongy mesophyll

30. In the leaf, _________ mesophyll is involved in gas exchange while __________ mesophyll is the primary photosynthetic tissue.

31. Companion and sieve tube cells are found in what transport tissue?

32. ___________ may be found on the upper or lower surface of leaves, or even on the stems of some plants. the size of this opening is controlled by __________ cells.

33. Simply put, wood is ________ ________ (two words).

34. Leaves attach to the stem at _______ (the regions in between these are called __________) by means of a stalk, the ________. The remainder of the leaf is called the _________.

For questions 35–42, match the effects with the correct hormones: auxins, cytokinins, gibberellins, ethylene, abscisic acid (sometimes called abscision acid).

35. Photo- and gravitropism.

36. Suppression of lateral (axillary) buds.

37. Elongation of cells.

38. Stimulates cell division, growth of lateral buds.

39. Causes bolting, affects fruit set.

40. Fruit ripening.

41. Opening and closing of stomata.

42. Prepares plant for winter dormancy.

43. Name (i) a plant hormone that promotes growth and (ii) one that retards or inhibits growth.

44. Seed germination is dependent upon the ratio of ____________ to _____________.

45. As __________ vapor is lost during transpiration, the water __________ in the leaf ________ (drops, rises) causing water to flow from the roots and the stem regions of __________ (higher, lower) water .

46. Carnivorous plants use their prey mainly as a source of ____________.

47. The closing of a Venus fly trap's leaf on an insect is an example of thigmotropism or response to _________________.

48. According to the pressure-flow hypothesis, sucrose moves in phloem because _________ moves into the phloem from the _________, creating a region of higher water __________.

49. Sieve tube elements and companion cells are part of the ___________ tissue, while vessel members (elements) and tracheids are found in ____________.

50. Match the cell type with the description.

Cell types: sieve tube elements, collenchyma, sclerenchyma, epidermal, endodermis, parenchyma, vessel members or elements, tracheids.

Descriptions:

(i) Cells with secondary cell walls reinforced by lignin, which lends rigidity to the plant, and are found in the ground tissue of stems. Mature cells have no cytoplasm or nucleus. Fibers, one of two types of this cell, are long and familiar from plant products such as rope (hemp, sisal, jute). The other type, schlerids, are shorter, more irregular cells and make up tough structures such as are found in seeds and nuts, and give a pear its characteristic texture.

(ii) These cells are reinforced at corners although they lack a secondary cell wall. These cells are found as support tissues just under the epidermis in newer stems or branches. These can form long, tough strands like those found in a stalk of celery.

(iii) Cells are unspecialized meristematic cells, and can differentiate into other types of cells. These cells are the most common in a vascular plant, and carry out the chores of being a plant: photosynthesis, respiration, metabolism, and storage. The generalized plant cell which is presented as a model is a parenchymal cell. These cells make up most of the ground tissue in a plant.

(iv) These cells form the phloem; they are responsible for the conduction of carbohydrates such as sucrose from the sites of photosynthesis (the sources) to the places where the sugars are used (the sinks). Anuclear at maturity, they depend on the closely-associated companion cells for nuclear activities. Both cells in a pair have a common parent cell.

(v) This is one of the cell types that make up the water-conducting tissue, xylem. Several cells can fuse and form long pipes.

(vi) Water moves from one to another of these cells, which make up the conducting pipe through areas in the cell walls that lack a secondary cell wall. Water must move through the cell walls and the middle lamella to be transported. At maturity these are dead.

(vii) These cells cover the exterior of the plant, protecting it from trauma and excessive loss of water. Guard cells that surround each stomate are epidermal cells which regular gas and water vapor exchange in leaves and stems. Root hair cells are specialized forms of this type of cell which function in nutrient uptake on roots. In a manner like the villi found in the intestine, these cells increase the surface area of roots with their very tiny finger-like projections.

(viii) These cells are common in roots, where waterproofed by suberin in their walls, they form a gasket-like barrier and allow selective transport of materials from the soil into the root. This waxy Casparian strip is actually inside the endodermal cell walls, and force water and solutes to move through cell plasma membranes and not seep through uncontrolled between cells.

51. Which vascular tissue is being described in each of the statements below? The choices (all used and some are used more than once) are: cambium, xylem, phloem, pith.

    a. _________which transports carbohydrate-rich liquid through the roots and stem, and which has parenchyma, sclerenchyma, sieve tube, and companion cells.

    b. __________ (includes parenchyma, sclerenchyma, vessel members, tracheid cells) which transport water and dissolved substances through the roots and stem.

    c. ________is a stem tissue composed of sclerenchyma and parenchyma cells and stores food.

    d. The secondary ________ and ________ is formed of vascular cambium (a meristematic region of actively dividing cells located between the xylem and phloem). Cork __________ is found outside the phloem and under the bark. Cork cells were observed by Robert Hooke in the 17th century when he applied the word "cell" to them.

52. a. List three leaf tissues and briefly describe the functions of each.

    b. Label the following diagram of a leaf in cross-section.

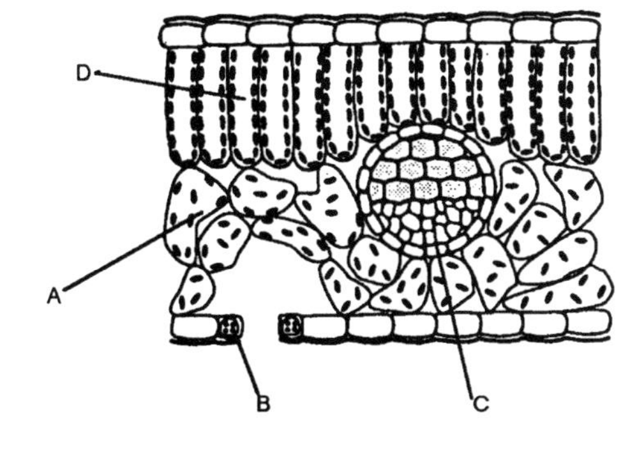

53. What is ground tissue?

54. What plant organ is being described in each of these passages?

    (i) ________ offer increased surface area for gathering the energy of sunlight for photosynthesis. ________ attach to the stem at nodes (the regions in between these nodes are internodes) by means of a stalk, the petiole. The remainder of the _____ is called the blade. This most closely describes dicot ______; monocot ________ usually have only a blade. _______ contain the tissues mentioned above, plus xylem and phloem in the leaf's veins. _______ come in many forms, e.g., spines on cacti, thick and water-filled in succulent plants, the tendrils of some climbing plants, or modified for food storage as in the onion (which is actually a large bud). When celery is consumed, it is the long petiole which is munched on.

    (ii) The ______ includes the stem and all that develops from it. Cells differentiate to perform special functions—into epidermis, phloem, xylem, and ground tissue. In the tip resides apical meristem. The ______ grows in length and the apical meristem moves with the tip. Along the stem another meristem, the axillary buds, will be left at points where leaves will develop. In some plants, the axillary bud meristem is inhibited from growing as long as the apical meristem remains functional. In others, the buds will become flowers or branches. Stems contain xylem and phloem and are the route for transportation of materials to and from the leaves and roots.

    (iii) _________ are specialized organs which actively and passively absorb water, gases, and nutrients from the soil and anchor the plant to its substrate. The outer region of the _____, the first concentric ring encountered as one moves from the exterior in, is the epidermis which contains very large numbers of _____ hairs. ______ hairs, which are individual cells with projections which may be several millimeters long, greatly augment the absorptive surface area of the root.

55. (i) Describe the anatomy of a typical root.

 (ii) Which part produces pollen?

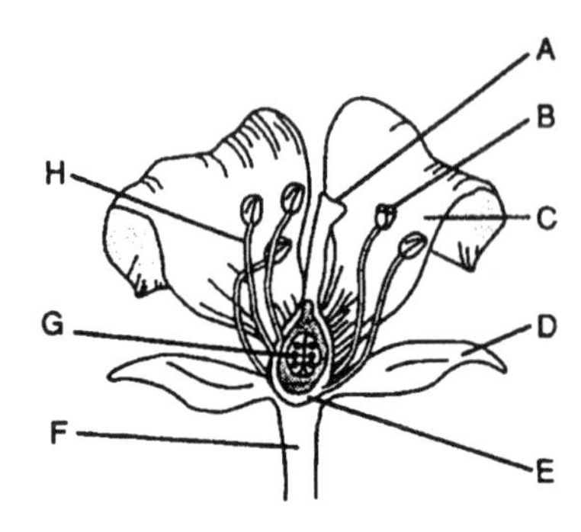

 (iii) Which part receives pollen in the process of fertilization?

 (iv) What is the structure indicated by "D" called?

56. Fill in the blanks with the name of the type of root being described:

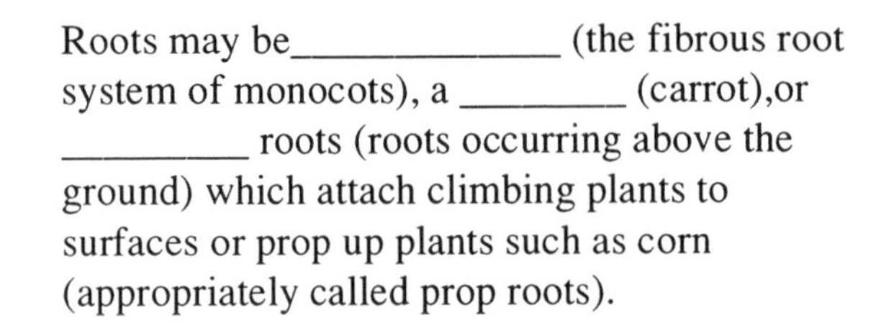
Roots may be_____________ (the fibrous root system of monocots), a ________ (carrot),or _________ roots (roots occurring above the ground) which attach climbing plants to surfaces or prop up plants such as corn (appropriately called prop roots).

57. How does a plant obtain water from the soil?

58. Describe plant responses to light, gravity, and contact.

59. Carnivorous plants are usually found in areas of ______ soil nitrogen.

 a. low

 b. no

 c. high

 d. very high

60. What are macronutrients and how are they defined? What are micronutrients?

Check Yourself

1. Roots absorb water, absorb nutrients and gases, and anchor the plant. Some roots may function as storage organs. **(Leaves, shoots, roots, and flowers)**

2. b. auxin distribution **(Plant hormones)**

3. Carbon, nitrogen, oxygen, phosphorus **(Water and nutrients)**

4. iron, copper, boron, zinc **(Water and nutrients)**

5. b. cyhtokinins and c. auxins **(Plant hormones)**

6. potassium ion, chloride ion. Hydrogen ions (protons) are pumped OUT of the guard cell in many or perhaps all plants. **(Water and nutrients)**

7. Plants lose water as water vapor by the process of transpiration. The stomata open, allowing for carbon dioxide to diffuse in, and this allows water vapor to diffuse out. Over 90 percent of a plants water is lost by transpiration. Guard cells surrounding the stomatal opening control water loss; guard cell turgor (a cell is turgid when it is plump with water) is controlled by actively pumping potassium ions into the cell. Chloride anions often follow the potassium ion into the cell, and hydrogen ions may be pumped out, maintaining electrical neutrality across the cell membrane. Guard cells are responsive to carbon dioxide concentration in that high concentrations cause them to close, and low concentration stimulate opening. Temperature and light also affect the guard cells. **(Water and nutrients)**

8. sucrose. **(Water and nutrients)**

9. Mineral elements are taken up in solution by plants, which is why soil acidity is important to farmers and gardeners. Soil pH affects the solubility of minerals. **(Water and nutrients)**

10. sources, sinks **(Water and nutrients)**

11. The water holding capacity of soil is field capacity. The type, amount, and size of soil particles (clay, sand, humus) determines field capacity. **(Water and nutrients)**

12. *Rhizobium*. Plant nutrition is a complicated topic. Some plants depend on symbiotic relationships with bacteria (*Rhizobium*) to fix atmospheric nitrogen in a form available to plants. In other plants, mycorrhizae, a fungus, absorbs nutrients from soil—including poor soils—which are then available to the plant. **(Water and nutrients)**

13. positively, gravitrophic **(Plant Responses)**

14. Etylene and auxin. When abscisic acid was discovered, it was thought to be responsible for leaf abscission, but no definite role has been determined. Changing concentrations of etylene and auxin causes the production of enzymes that digest cellulose and other components of cells walls. **(Plant Hormones)**

15. heartwood **(Wood)**

16. leaves **(Leaves, shoots, roots, and flowers)**

17. root hairs, epidermis, cortex, pericycle, vascular cylinder, pith **(Leaves, shoots, roots, and flowers)**

18. In germination, favorable environmental conditions rouses the embryo from its dormant state, causing it to enter a period of rapid growth and development and the formation of the root and stem. A seed embryo has three parts: (i) cotyledon(s) or storage leaves which wither once their food stores are used up—one cotyledon in monocots, two in dicots; (ii) the hypocotyl which becomes the lower stem and root of the plant; and (iii) the epicotyl, which produces the upper stem and leaves. The conditions needed to stimulate germination varies widely among plants; some need a warm period, others a cold period or even freezing or exposure to fire or to light of the right wavelength for the right length of time. Conditions providing moisture and warmth will induce many seeds to germinate. **(Germination)**

19. girth (or diameter). Secondary growth refers to increases in diameter (or girth). It is not seen to more than a very limited extent in herbaceous plants (including most monocots). Lateral meristematic tissue, the vascular cambium (lies between the xylem and phloem) and cork (in the cortex) cambiums, form secondary tissues, secondary xylem and secondary phloem among others. Secondary xylem is located closest to the center of the plant, secondary phloem toward the outside. **(Growth)**

20. The apical meristem produces auxin which inhibits the growth of the lateral (axillary) buds. **(Plant Hormones)**

21. sclerenchyma **(Cells and tissues)**

22. parenchyma **(Cells and tissues)**

23. sclerenchyma, specifically schlerid cells **(Cells and tissues)**

24. epidermis **(Cells and tissues)**

25. sclerenchyma **(Cells and tissues)**

26. endodermis. Most of a root is cortex. Here cells may be modified for storage as in carrots. The innermost portion of the cortex is the endodermis which forms a tight band made waterproof by the waxy Casparian strip, which acts as a gasket. At this point, movement of water and solutes which moved freely in the cortex can only move through the cell membranes of endodermal cells, providing a control point for regulating uptake of these substances. **(Cells and tissues)**

27. True **(Cells and tissues)**

28. a. gymnosperms and c. ferns. Angiosperms have tracheids and vessel element cells. Bryophytes do not have conducting tissues in that they are non-vascular. **(Cells and tissues)**

29. c. Casparian strip **(Cells and tissues)**

30. spongy, palisade **(Leaves, shoots, roots, and flowers)**

31. phloem **(Cells and tissues)**

32. stomata, guard **(Leaves, shoots, roots, and flowers)**

33. secondary xylem. Wood is secondary xylem. The inner layer of wood is called heartwood and contains a high proportion of resin. It no longer functions in water transport. The outer sapwood is still functional. Annual tree rings result from the alternation of early (spring) and late (summer) woody tissue. In spring the xylem cells are mostly larger vessel cells, giving an open appearance, while the late wood, composed of smaller numbers of vessel cells and tracheids, appears denser. Trees that grow in regions where the seasons vary but little (such as the tropics) do not go through distinct growing seasons and cannot be reliably dated by counting tree rings. **(Wood)**

34. nodes, internodes, petiole, blade. **(Leaves, shoots, roots, and flowers)**

35. auxins (causes cells elongation on side away from light, for instance, bending the stem toward the light source). **(Plant hormones)**

36. auxins (apical dominance effect). **(Plant hormones)**

37. auxins **(Plant hormones)**

38. cytokinins **(Plant hormones)**

39. gibberellins **(Plant hormones)**

40. ethylene **(Plant hormones)**

41. abscisic acid **(Plant hormones)**

42. abscisic acid **(Plant hormones)**

43. (i) auxins, cytokinins, and gibberellins promote growth; (ii) ethylene and abscisic acid inhibit growth. Plant hormones are of five types:

    (1) Auxins which mainly cause cells to elongate, but also promotes root growth, branching, apical dominance, and fruiting. Gravitropism (response of a plant to gravity) is affected by auxins, as is phototropism. After synthesis by the shoot apical meristem, auxins move downward going from cell to cell. Auxins contain indoleacetic acid (IAA). Auxins generally promote growth.

(2) Cytokinins in angiosperms are mainly produced in roots, stimulate cell division, cause development of lateral buds and deter root formation. Cytokinins and auxins are antagonists. The balance of the levels of each in the plant body determines, e.g., whether axillary bud growth will occur or not. Cytokinins are found in a number of organisms and generally promote growth.

(3) Gibberillins are made in the shoot apex and root apex, young leaves and in the embryo. They are responsible for bolting (where a stem suddenly grows longer as in spinach at the beginning of the summer), fruit set, flowering, fruit development, and seed germination (breaks dormancy). Gibberellins generally promote growth.

(4) Many years ago, it was noted that trees near gas street lamps lost their leaves due to the presence of ethylene in the gas. Ethylene is a fairly simple gas which inhibits root and shoot growth and lateral bud production while favoring the dropping of fruit and leaves (abcission). Leaves drop from plants as less daylight, cooler temperatures reduce the levels of auxin in the plant, and this, in turn, increases the production of ethylene. Ethylene stimulated the making of enzymes which can digest cell walls, weakening the attachment of leaves and fruit to the plant. Apples give off ethylene gas while ripening, which is why they can speed up the ripening of nearby bananas. Ethylene generally inhibits growth. Potatoes can be prevented from sprouting by placing an apple and the potatoes in a plastic bag or keeper drawer of the refrigerator.

(5) Abscisic acid retards growth, causes leaves to age, controls the stomata (and other aspects of plant physiology related to water stress), and is involved in preparing the plant for winter dormancy. It also inhibits both primary and secondary growth and causes dormancy in the apical buds and in seeds. Abscisic acid generally inhibits growth.

A key point in understanding plant hormones is to realize they act in concert. For instance, the interplay auxin, cytokinins and ethylene determines axial bud growth. Seed germination is caused or prevented depending on the ratio of abscisic acid to gibberellins. **(Plant hormones)**

44. abscisic acid, gibberellin **(Germination)**

45. water, potential, drops, higher, potential. Water movement is often discussed in terms of water potential. The direction in which water flows is simply as would be expected, from a region of higher water potential to a region of lower water potential. Transpirational water loss from leaves reduces the water potential in the leaf; a higher water potential exists in the roots and xylem of the stem. This leads to a net flow of water and solutes from the roots to the leaves. **(Water and nutrients)**

46. nitrogen **(Water and nutrients)**

47. touch **(Plant Responses)**

48. water, xylem, potential. Explanation: Carbohydrates and other materials are transported in plants in phloem, from the regions where they are made (the sources) to areas of storage or use (sinks). The process of transport is known as translocation. Differences in solute concentration result in osmosis and the passive transportation of free sucrose. Energy is only required to move sucrose into or out of the phloem. According to the pressure-flow hypothesis, sucrose is pumped at the source into the phloem, which causes water from the xylem to move into the phloem as well (osmotic pressure). At the sink, the sucrose is actively pumped out of the phloem. This decreases the concentration of sucrose, and water flows out back into the xylem by osmosis. The pressure in this circuit drives the flow of translocation. The process can be best understood in terms of relative water potential. **(Water and nutrients)**

49. phloem, xylem. **(Cells and tissues)**

50. (i) Sclerenchyma cells (ii) Collenchyma cells (iii) Parenchyma cells (iii) Parenchyma cells (iv) Sieve tube elements or members. (v) vessel members or elements (vi) tracheids (vii) epidermal cells (vi) endodermis cells. **(Cells and tissues)**

51. a. phloem, b. xylem, c. pith, d. xylem, phloem, cambium. **(Cells and tissues)**

52. a. Leaf tissues: (1) Spongy mesophyll involved in gas exchange and photosynthesis. (2) Palisade mesophyll, the primary photosynthetic tissue. (3) Guard cells (also found on stems) can be found on the lower and/or upper leaf surfaces and regulate the stomatal opening allowing gas exchange by diffusion but restricting water loss. (b) A is spongy mesophyll, B is a guard cell, C is the vascular bundle, and D is palisade mesophyll. **(Cells and tissues)**

53. Ground tissue makes up most of the main part of the plant. It is composed of parenchyma, sclerenchyma and collenchyma cells. Permanent tissues which are not dermal or vascular are considered ground tissue. **(Cells and tissues)**

54. (i) leave or leaves, (ii) shoot, (iii) root or roots. **(Leaves, shoots, roots, and flowers)**

55. (i) A dicot root may serve as an example of root anatomy. At the root tip is a root cap, which is a protective layer of cells constantly replenished as they wear away when the root pushes throughout the soil. Behind this lies the apical meristem from which arises the root cap and the various differentiated cells and tissues of the root. Above a region of maximum cell multiplication is the zone of elongation. Here cells grow to a number of times their original length, pushing the root into the soil. Above this level, the cells differentiate into primary meristems which become the root tissues: primary xylem and phloem, cortex, and epidermis. This is called the zone of differentiation. As can be deduced from these facts, root hairs are not found below this zone.

    (ii) letter B, the anther

    (iii) letter A, the stigma

    (iv) sepal **(Leaves, shoots, roots, and flowers)**

56. Adventitious, taproot, aerial **(Leaves, shoots, roots, and flowers)**

57. Plants obtain water from the soil by the following mechanisms. Root cells have a higher concentration of solutes than is present in the soil, and water therefore moves into the roots by osmosis. This creates root pressure, which is only strong enough to move the water a short way up the stem. Transpiration at the top of the plant supplies the rest of the motive force needed to move water up to the leaves. Water in the vascular tissues forms a column from the roots to the leaves. As water molecules diffuse out, a tug is put on this column, since all the water molecules are joined by hydrogen bonds (cohesion). The pull is transmitted along the column and the water pulled up. In part, the column exists due to capillary adhesion between the water and the sides of the narrow conducting tubes. Ultimately, movement of water through the plant is a passive process powered by the sun's energy (causing transpiration) and made possible by hydrogen bonding between water molecules. **(Water and nutrients)**

58. Plants respond to a number of stimuli. Trophisms are movements due to growth in which a plant responds to a particular stimulus. In phototropism (light stimulation), roots are negative (bend away from light) and shoots are positive (bend toward light) orienting the plant. Auxins increase in concentration on the darker side (away from the light) fostering increased cell elongation, and bend the shoot toward the light. Gravitropism (response to gravity) is not as well understood as phototropism. Roots are positively gravitrophic (grow downward) and shoots are negatively gravitrophic (grow upward). Amyloplasts (statoliths) containing starch accumulate on the lowest side of a cell. Where they settle, gibberellins and

auxins are found in higher concentrations. They stimulate growth, causing a bending upward. ("Gravitropism" was formally called "geotropism.") Thigmotropism is growth which results from contact, as seen in vines. No generally accepted explanation has been put forth for this behavior.

Plants exhibit a number of other responses as well—guard cells are sensitive to certain wavelengths of light, as well as carbon dioxide concentrations. Water stress induces wilting, which reduces the leaf surface area exposed to the sun. **(Plant responses)**

59. a. low. Carnivorous plants, like the Venus fly trap and pitcher plant, attract and trap insects as a source of nitrogen. They are usually found growing in areas of low soil nitrogen. **(Water and nutrients)**

60. The macronutrients included nitrogen, carbon, oxygen, phosphorus, calcium, sulfur and magnesium. Micronutrients are found and required in lesser amounts, and include iron, boron, copper, zinc, etc. Major chemical elements in fertilizers are nitrogen, phosphorus, and potassium. The numbers on packages of fertilizer (such as "14 - 8 - 6") refer to the percentages the fertilizer contains of those elements, in that order. Some fertilizers also contain other macro- and micronutrients. **(Water and nutrients)**

Grade Yourself

Circle the question numbers that you had incorrect. Then indicate the number of questions you missed. If you answered more than three questions incorrectly, you will then have to focus on that topic. (If a topic has less than three questions and you had at least one wrong, we suggest you study that topic also. Read your textbook, a review book, or ask your teacher for help.)

Subject: Plants: Structure, Growth, Development, and Physiology

Topic	Question Numbers	Number Incorrect
Leaves, shoots, roots, and flowers	1, 16, 17, 30, 32, 34, 54, 55, 56	
Plant hormones	2, 5, 14, 20, 35, 36, 37, 38, 39, 40, 41, 42, 43	
Water and nutrients	3, 4, 6, 7, 8, 9, 10, 11, 12, 45, 46, 48, 57, 59, 60	
Plant responses	13, 47, 58	
Wood	15, 33	
Germination	18, 44	
Growth	19	
Cells and tissues	21, 22, 23, 24, 25, 26, 27, 28, 29, 31, 49, 50, 51, 52, 53	

Digestive, Immune, and Endocrine Systems

Brief Yourself

Nutrition

An organism's diet provides it with energy from carbohydrates, fats, and proteins, as well as other nutritional factors such as vitamins and minerals. Vitamins and minerals have diverse functions, acting in concert with enzymes and even forming tissues in bone and teeth.

Digestion

Digestion is the process of breaking down materials to make simpler compounds the body can assimilate. Digestion can occur outside the body, e.g., fungi secrete enzymes and digest food outside their cells, absorbing nutrients when digestion is complete. Generally, animals take food into their bodies to digest and assimilate. A single-celled protist does this by the process of phagocytosis, bringing the food material into a vacuole within the cell and digesting it there. In higher organisms, the animal's digestive tract processes food in a step-by-step manner.

Cell-mediated and humoral immunity

The immune system is responsible for defending the body against foreign invaders such as bacteria and viruses. The immune system recognizes foreign proteins (and some polysaccharides) called antigens, present on the surfaces of microorganisms and foreign cells, and mounts a defensive reaction. It fights the invasion with antibodies and various white blood cells. The antibodies attach to the antigens marking cells and causing them to be more readily consumed by white cells like macrophages. After one infection by a specific microorganism, antigens related to it are "on file," and the body typically reacts to it very quickly in subsequent encounters, often minimizing or preventing later infections. We get repeated infections like colds because slightly different viruses attack us each time, or in some cases the same virus evolves quickly. Either way, the earlier antigen is no longer present in the protein coat of the virus.

Hormones

Hormones are chemical messengers produced by glands that secrete directly into the blood stream (endocrine or "ductless" glands), and perfom regulatory functions in the body. They are peptides, proteins, or steroids, and effective in very low concentrations. Hormones affect a specific body site called the target tissue. For instance, epinephrine is a hormone produced in the medulla (center) of the adrenal gland. While it is released into the general circulation and has multiple effects, the effects are each due to interaction with a target tissue, such as in the heart's pacemaker, where epinephrine increases heart rate. Hormones generally control a cell by influencing the production of enzymes or enzyme catalyzed reactions. A molecule commonly involved in hormone action is cylic AMP. Important glands within the endocrine system include the pituitary gland and hypothalamus in the brain, the adrenal gland (really two glands in one), the ovaries and testes which secrete sex hormones, the thyroid (produces thyroxine and affects metabolism), parathyroid (affects calcium uptake and storage) and pancreas (makes insulin/glucagon).

Test Yourself

1. Vitamins for the most part function as

   a. coenzymes

   b. cofactors

   c. catalysts

   d. sources of energy

2. True or false; if false, correct to make the statement true. Animals can only use carbohydrates and fats as sources of energy, not proteins.

3. Of the 20 amino acids, how many are essential for humans, that is, must be obtained from the diet?

4. Why do birds and mammals have higher basal metabolic rates than other vertebrates?

5. Match the mode or form of absorption with the nutrient:

   Nutrient:

   (i) carbohydrates

   (ii) proteins

   (iii) most fatty acids and glycerol

   Mode or form:

   a. as monosaccharides by active transport and facilitated diffusion

   b. as disaccharidase by active transport alone

   c. as amino acids by active transport

   d. as amino acids by passive diffusion

   e. enter cells lining the intestine by passive diffusion, and re-synthesized into fats

   f. enter cells lining the intestine by active transport and re-synthesized into fats

6. The main source of vitamin K are ____________ in the _______ intestine.

7. What is a ruminant?

8. Some humans have discomfort after consuming milk products because they lack the enzyme for digesting the __________ in milk.

9. What is a cecum? What are some animals that have a large cecum?

10. How are termites able to digest cellulose?

    a. They produce the necessary digestive enzymes

    b. Cellulose is easy for most animals, including termites, to digest

    c. Termites can't digest cellulose

    d. Protists in the termite gut digest the cellulose

11. List the path food takes through a person's digestive tract. Then briefly describe digestion and absorption at each stage of the passage through the gut.

12. List the enzymes that are secreted in the

    (i) mouth

    (ii) stomach

    (iii) small intestine

13. When swallowing a drink of water or some food, what prevents the material from entering the trachea?

14. Digested materials enter the villi in the small intestine and are transported to the _____ via the ________ system. Fatty material absorbed by the lacteals enters the _________ system from the villi and is eventually dumped into the __________.

15. The progressive contraction of the digestive tract that move food along is called

    a. peristalsis

    b. rumination

    c. diarrhea

    d. deglutition

16. The liver performs several functions in digestion. List two.

17. Enzymes produced in the pancreas enter the _____________ by way of the ___________.

    a. large intestine, bile duct

    b. jejunum, pancreatic duct

    c. duodenum, pancreatic duct

    d. stomach, bile duct

18. What viral disease was eliminated from the earth in the 1970s through the efforts of the World Health Organization?

19. What is inflammation? What chemical is released (among others) by injured cells that can trigger the inflammatory response?

20. What is the location of and type of cell responsible for producing white blood cells?

    a. long bones, macrophages

    b. long bones, stem cells

    c. ribs, erythrocytes

    d. ribs, lymphocytes

Questions 21–29. Matching.

Possible answers:

a. Antibodies involved in allergy

b. Enzyme present in tears and sweat that attacks bacterial cell walls

c. Enlarged section of the lymph system inhabited by macrophages

d. Lymphocytes—produce antibodies active against bacteria and viruses

e. Large, complex, globular protein molecules which react against a specific antigen

f. Classes of antibodies or immunoglobulins

g. Molecule capable of provoking an immune response

h. Lymphocytes—mainly active against eukaryotic cells, in cell-mediated immunity

i. Phagocytic cells found in lymph nodes, liver, spleen

21. Macrophage

22. Lymph nodes

23. B cell

24. T cell

25. IgE

26. IgE, IgM, IgA, IgD, IgE

27. Lysozyme

28. Antibodies

29. Antigen

30. Give two examples of non-specific host defenses, that is, defenses against infection that do not involve humoral or cell-mediated immunity.

31. What are hormones?

32. Name a hormone that has the following action, and give the source of the hormone.
    (i) Regulates blood sugar level (causes it to fall)
    (ii) Increases heart rate
    (iii) Stimulates production of milk in mammals
    (iv) A protein that affects salt and water balance
    (v) Affects metabolic rate
    (vi) Regulates blood sugar level—causes it to rise
    (vii) Beard growth and deeper voice in males
33. What hormone is responsible for molting in insects?
    a. Insulin
    b. Ecdysone
    c. Juvenile hormone
    d. ACTH
34. Hormones used for communication with other members of a species are called ___________.
35. __________ _________ (two words) maintains insects in the larval stage.
36. Write the names for the following abbreviations. Where do these hormones originate?
    (i) FSH
    (ii) TSH
    (iii) LH
    (iv) ACTH
37. Describe functions for two of the hormones listed in question 36.
38. ___________ are hormones produced by the follicle of the ovary; the corpus luteum produces these as well as ____________, the hormone of pregnancy.
39. What is the action of the parathyroid gland?
40. Name one hormone produced in the hypothalamus, and briefly describe its function.
41. The adrenal _________ is located at the center of the adrenal gland and secretes two important hormones, ___________ and _____________.
42. Prostaglandins have hormone-like actions, but differ from hormones in the which of the following ways? (Pick all that apply.)
    (i) They are proteins
    (ii) They are fatty acids
    (iii) Prostaglandins are produce in most or all tissues of the body
    (iv) They are produced and used locally
    (v) Prostaglandins are produced by the prostate
43. Distinguish between endocrine and exocrine.
44. Hormones act by means of receptors. Match the hormones with the receptor.

    Receptors:
    (i) Located on the plasma membrane
    (ii) Located in the cytoplasm or nucleus

    Hormones: thyroid and steroid hormones, protein and peptide hormones
45. A hormone binding with a cell receptor often releases a secondary chemical message, which for many hormones is known as cyclic _____. (abbreviation)
46. Why can thyroid and steroid hormones freely diffuse into cells, across the plasma membrane? Why is this important?
47. Another name for somatotropin is _________ __________. (two words)
48. Discuss how the digestive tract varies among animals.
49. Briefly discuss the functions of the pancreas.
50. Compare humoral and cell-mediated immunity, including a list of the classes of antibodies, and the roles of T- and B-cells.

51. In addition to the agents of immunity mentioned in question 50, what other defenses do humans have against infection?

52. What two parts of the vertebrate brain have important functions as part of the endocrine system? What roles do they play?

53. What is the physiological significance of the two parts of the adrenal gland, the medulla, and cortex?

54. How can the chemical structure of a hormone molecule give clues to the way it acts in cells?

Check Yourself

1. a. coenzymes **(Nutrition)**

2. False. Animals can use carbohydrates, fats, and proteins as sources of energy. **(Nutrition)**

3. 8 of the 20 amino acids are essential and must be obtained form the diet. **(Nutrition)**

4. Birds and mammals are endothermic, which means that they main a fairly constant body temperature, and their metabolic rate reflects this additional demand. **(Nutrition)**

5. (i) a

   (ii) c

   (iii) e **(Digestion)**

6. bacteria, large **(Digestion)**

7. Ruminants are herbivores such as cattle and deer who possess a complex digestive system. Ruminants have cellulose-digesting bacteria and protozoans (protists) in their digestive tract. Plant materials are first digested in the rumen (the first of the four "stomachs" of the cow), and then in the reticulum. Undigested material is brought up back into the mouth for further chewing ("chewing the cud"). The material is swallowed again. Much of the energy from the plants eaten by ruminants is obtained from fatty acids produced by digestive tract bacteria. **(Digestion)**

8. Lactose or milk-sugar, a disaccharide **(Digestion)**

9. The cecum or blind gut is a pouch which looks much like a thumb and projects off the end of the small intestine, where it joins the large intestine. Ceca are found in many mammals such as horses, rabbits and rats. In the cecum bacteria further digest plant material, enabling the animal to get more nutrients out of its food. **(Digestion)**

10. d. protists in the termite gut digest the cellulose. **(Gastrointestinal tract)**

11. Mouth-pharynx-esophagus-stomach-pyloric valve-duodenum-small intestine-large intestine-rectum-anus.

    In humans, food is broken up by chewing, and mixed with saliva which contains salivary amylase, an enzyme which converts starch, primarily, into simpler sugars. Passing to the stomach through the esophagus, the food is further broken down by the action of hydrochloric acid and the protein-digesting enzyme, pepsin. Passing into the small intestine, fats are broken into droplets by bile produced in the liver, and broken into fatty acids and glycerol by lipases. Proteins are broken apart by one type of proteolytic enzyme, reduced to dipeptide fragments by a second type of enzyme and finally split into individual amino acids by a third enzyme type. Amylases made in the pancreas continue the work breaking down starches. Disaccharidase are finally broken into monosaccharides by enzymes in the membranes of the cells lining the intestine (epithelial cells) Absorption occurs in a manner specific to the type of nutrient. Monosaccharides are absorbed by active transport and facilitated diffusion, and amino acids and dipeptides are moved by active transport. Large fatty acids and glycerol diffuse into the epithelial cells, and are combined into new fat molecules; these coalesce into droplets which are coated with proteins. The droplets enter the lymph system and are finally dumped into the blood stream when lymph is returned to the general circulation. A special form of circulation exists linking capillaries in the intestine with capillaries in the liver, the hepatic portal system. This allows substances, especially monosaccharides, to move directly from the site of absorption to the liver for processing. Water, ions, and minerals are absorbed in both small and large intestines. Disruption of the absorption of water, due to bacteria or dietary factors, results in diarrhea. The remaining materials—indigestible plant fibers, bacteria, other materials—are passed from the body as feces. **(Gastrointestinal tract)**

12. (i) salivary amylase (ptyalin)

    (ii) pepsin

    (iii) pancreatic amylase, trypsin, chymotrypsin and others **(Digestion)**

13. Swallowing is a coordinated act in which the epiglottis (a flap of tissue at the back of the throat) blocks the trachea. **(Gastrointestinal tract)**

14. liver, portal, lymphatic, bloodstream **(Gastrointestinal tract)**

15. a. peristalsis **(Gastrointestinal tract)**

16. The liver produces bile, a natural detergent which emulsifies fats, making them more digestible. The liver also metabolizes and processes many materials absorbed in digestion, including detoxifying many otherwise toxic compounds, and stores glucose in the form of the polymer glycogen (animal starch). **(Gastrointestinal tract)**

17. c. duodenum, pancreatic duct **(Gastrointestinal tract)**

18. Smallpox **(Other defenses against infection)**

19. Inflammation is a reaction to injury that includes localized swelling, pain, redness, increased temperature, and often temporary loss of function. It is a protective reaction, the first step toward defending against potential invading microorganisms. Histamine is one of a number of chemical released by damaged tissues. **(Other defenses against infection)**

20. b. long bones, stem cells **(Cell-mediated and humoral immunity)**

21. i. phagocytic cells found in lymph noes, liver, spleen **(Cell-mediated and humoral immunity)**

22. c. enlarged section of the lymph system inhabited by macrophages **(Cell-mediated and humoral immunity)**

23. d. lymphocytes **(Cell-mediated and humoral immunity)**

24. h. lymphocytes. **(Cell-mediated and humoral immunity)**

25. a. antibodies involved in allergy **(Cell-mediated and humoral immunity)**

26. f. classes of antibodies **(Cell-mediated and humoral immunity)**

27. b. enzyme present in tears **(Other defenses against infection)**

28. e. large complex, globular protein molecules **(Cell-mediated and humoral immunity)**

29. g. molecule capable of provoking an immune response **(Cell-mediated and humoral immunity)**

30. (i) The skin acts as a barrier to microorganisms.

    (ii) The skin has an acid pH which inhibits the growth of many bacteria. Other examples could include the hairs, cilia and mucus produced in the respiratory tract, the normally acid pH of the vagina, lysozyme in tears and sweat. **(Other defenses against infection)**

31. Hormones are biochemical compounds produced in one part of the body that are capable in minute doses of affecting a different part of the body. Hormones are chemical messengers that affect a target tissue and are produced by ductless glands. They may be a steroid (a class of lipids) or a protein (or peptide). **(Hormones)**

32. (i) insulin, islets of Langerhans in the pancreas, beta (β) cells

    (ii) epinephrine (adrenaline) produced in the adrenal medulla

    (iii) prolactin, anterior lobe of pituitary gland

    (iv) antidiuretic hormone (ADH), hypothalamus. Aldosterone has a similar effect, but as suggested by its name, is a steroid and secreted by the adrenal cortex.

    (v) thyroxine, thyroid gland

    (vi) glucagon, islets of Langerhans in pancreas, the alpha (α) cells

    (vii) testosterone, testes **(Hormones)**

33. b. ecdysone **(Hormones of invertebrates)**

34. pheromones **(Hormones)**

35. Juvenile hormone. Special aspects of the life cycles of many invertebrate animals are controlled by hormones. All insects that metamorphose and are maintained in the larval form by juvenile hormone, and ecdysone, produced by the prothoracic gland, induces growth, molting and development into the adult form. Insects (and many other animals) produce unique "social hormones" called pheromones, which affect behavior of other individuals, such as attracting potential mates. **(Hormones of invertebrates)**

36. (i) FSH stands for follicle-stimulating hormone (anterior lobe of pituitary)

    (ii) TSH stands for thyroid-stimulating hormone (anterior lobe of pituitary)

    (iii) LH is luteinizing hormone (anterior lobe of pituitary)

    (iv) ACTH is adrenocorticotrophic hormone (anterior lobe of pituitary)

    ("Adrenocorticotrophic" is large word, and easy to stumble over, but just as it can be pronounced by breaking it down into parts, its meaning can be easily deciphered in that way, too: adreno- refers to the adrenal gland; -cortico- refers to the cortex of that gland; and -trophic refers to an action. Breaking long words like this one down can often help in recall.) **(Hormones)**

37. FSH stimulates the ovarian follicles in women, and sperm production (spermatogenesis) in men.

TSH stimulates the thyroid to release thyroxine.

LH induces the formation of corpora lutea (singular is corpus luteum) in females, and interstitial cells in the testes of males.

ACTH stimulate the adrenal cortex. **(Hormones)**

38. estrogens, progesterone **(Hormones)**

39. The parathyroid gland promotes the uptake of calcium by:

    (i) reducing the loss of calcium in urine,

    (ii) favoring the conversion of vitamin D (formed in our skin in sunlight) into its biologically active form

    (iii) by causing calcium to be liberated from bones. **(Hormones)**

40. Oxytocin is a hormone produced in the hypothalamus. It causes milk letdown (milk ejection) and can cause uterine contractions. Or, antidiuretic hormone that inhibits the formation of urine. **(Hormones)**

41. medulla, epinephrine (adrenaline), norepinephrine (noradrenaline) **(Hormones)**

42. Items ii, iii, iv are all true. Prostaglandins were originally associated with the prostate, hence their name. **(Hormones)**

43. Exocrine glands (such as sweat glands) use a duct to deliver their products, whereas endocrine glands are ductless, and secrete their products into the blood stream. **(Hormones)**

44. (i) protein and peptide hormone receptors are found in the plasma membrane

    (ii) thyroid and steroid hormone receptors are located in the interior of the cell.

    See question 46 below. **(Hormones)**

45. AMP (also called cAMP). **(Hormones)**

46. Thyroid and steroid hormones are freely soluble in fats and lipids. Thus, the plasma membrane doesn't constitute a barrier to them, and they are free to diffuse into the cell—to where their receptors are located. **(Hormones)**

47. growth hormone **(Hormones)**

48. The anatomy and physiology of the digestive tract varies among animals. Hydra have a simple cavity in which material is attacked by enzymes, more complex animals have a digestive tract broken down into specialized sections. Birds, for instance, have a crop in which food is held and a gizzard where food is "chewed" by the action of gravel churned by muscles in the gizzard wall. In mammals, there are specialized digestive tracts. Ruminants such as cows, deer and sheep, ingest fibrous fodder such as grass and digest it in a series of pouches loosely referred to as "stomachs." The food ("cud") is regurgitated into the mouth for further chewing during this process. Bacteria and protists break down otherwise indigestible substances such as cellulose, and make fatty acids that are absorbed by the host animals. Other animals also have assistance from microorganisms in digestion. In humans, vitamin K (needed for proper blood clotting) is made by bacteria in the gut. Animals as diverse as horses, rabbits, and rats have a specialized pouch off of the intestine, called the cecum. Here bacteria further digest foodstuffs, though no where as extensively as in ruminants. **(Gastrointestinal tract)**

49. The pancreas has several functions, each related to one of the several tissues that make up this organ. The islets of Langerhans make insulin, and glucagon, hormones important in the regulation of glucose use. Other parts of the pancreas make digestive enzymes. **(Hormones)**

50. Broadly speaking there are two closely interconnected divisions of the immune system: (i) Cell-mediated immunity which provides immunity to specific antigens through T cells (T-lymphocytes) and (ii) humoral immunity which involves the production of antibodies which are synthesized by B cells (B-lymphocytes). An antibody is a globular protein produced against a specific antigen. It is formed of two long and two short polypeptide chains, the amino acid sequence of which is for the most part constant from antibody to antibody, and has a pair of antigen-binding sites where the amino acid composition varies. There are five classes of antibodies (immunoglobulins) in mammals: Immunoglobulin A (IgA), IgM, IgG, IgD, and IgE. These classes are specialized in their functions and where they are found. IgA is found in secretions such as mucus and milk. Antibodies from B cells are active mainly against bacteria and viruses.

    T-cells have antibody-like receptors on their surfaces, and multiply when they recognize an antigen. T cells of different types have different roles: helper T cells recruit other white cells to attack invading organisms and stimulate the production of antibodies by B cells, cytotoxic (killer) T cells help kill cells that are infected with viruses, and suppressor T cells make T and B cells less active. T and B cells are produced in the bone marrow as undifferentiated stem cells; T cells migrate to and mature in the thymus (hence their name). Mature T- and B-cells are found in the spleen and lymph nodes. **(Cell-mediated and humoral immunity)**

51. Other defenses against infection include the acid pH of the skin and vagina, which retards the growth of most harmful bacteria and fungi. Tears and sweat contain an enzyme, lysozyme, which attacks the cell walls of bacteria. Other defenses include the physical barriers of skin and mucous membrane, acid in the stomach, constant production and outward movement of mucus in the respiratory tract, wax in the ear canal, and the presence of innocuous or beneficial bacteria in the gut, upper respiratory tract, and on the skin. **(Other defenses against infection)**

52. Two parts of the vertebrate brain are of special importance in the endocrine system the pituitary and hypothalamus. The pituitary gland, which is actually controlled by the hypothalamus, is divided into two parts, and makes a number of different peptide/protein hormones. The posterior pituitary secretes two hormones made in the hypothalamus: oxytocin and vasopressin (antidiuretic hormone or ADH). In the anterior pituitary, four tropic hormones (TSH, ACTH, FSH, and luteinizing hormone) are made. The anterior pituitary also secretes two non-trophic hormones, growth hormone and prolactin. (The anterior pituitary is sometimes called in popular literature the "master gland" since it affects other glands through these trophic hormones, but it must be remembered that the control of the anterior pituitary itself is actually in the hypothalamus.) **(Hormones)**

53. The adrenal glands are really two endocrine glands: the cortex (outer layer) which secretes corticosteroids and steroids, including glucocorticoids and mineralocorticoids, which affect blood sugar and water-salt balance, respectively. The adrenal medulla (inner part) secretes epinephrine and norepinephrine, which affect a variety of tissue and play a role in the so-called "fight-or-flight" response. **(Hormones)**

54. The chemical makeup of a hormone is a clue to where it acts within the cells: (i) steroid hormones like estrogen are a class of lipids, and as such pass through the plasma membrane, and alter gene expression in the nucleus of a cell; whereas (ii) peptide and protein hormones like insulin react with receptors on the cell surface, and the message is carried into the cell interior by a secondary molecule, or the plasma membrane's permeability is changed, allowing ions to enter the cell. **(Hormones)**

Grade Yourself

Circle the question numbers that you had incorrect. Then indicate the number of questions you missed. If you answered more than three questions incorrectly, you will then have to focus on that topic. (If a topic has less than three questions and you had at least one wrong, we suggest you study that topic also. Read your textbook, a review book, or ask your teacher for help.)

Subject: Digestive, Immune, and Endocrine Systems

Topic	Question Numbers	Number Incorrect
Nutrition	1, 2, 3, 4	
Digestion	5, 6, 7, 8, 9, 12	
Gastrointestinal tract	10, 11, 13, 14, 15, 16, 17, 48	
Other defenses against infection	18, 19, 27, 30, 51	
Cell-mediated and humoral immunity	20, 21, 22, 23, 24, 25, 26, 28, 29, 50	
Hormones	31, 32, 34, 36, 37, 38, 39, 40, 41, 42, 43, 44, 45, 46, 47, 49, 52, 53, 54	
Hormones of invertebrates	33, 35	

Nervous System and Senses

Brief Yourself

General

The nervous system is responsible for obtaining information about the external and internal environment in animals through the use of the various senses. Information is interpreted, integrated with other information (in some instances stored as memories) and action is taken. The nervous system includes senses, such as the familiar hearing, seeing, feeling pressure and heat or cold, as well as senses unfamiliar to humans such as lateral lines in amphibians and fish. Animals exhibit a full range of nervous system complexity, ranging from the relatively straightforward (as in insects) to the very sophisticated (as in the vertebrates).

Vertebrate Central Nervous System

Within the spinal cord, sensory neurons coming in from the dorsal side are interconnected with interneuron in the spinal cord; these transmit the signal to a motor neuron, which finally tells a muscle to contract. This is a simple reflex arc, and is accomplished within the spinal cord. This is what allows such rapid retraction of a hand from a hot object. Sensory information can also be transmitted up to the brain via relay neurons in the spinal cord.

Vertebrate Peripheral Nervous System

Axons from neurons in the PNS reach to the exterior and interior of the body. Sensory neurons collect and transmit information, while motor neurons carry signals out, causing muscles to contract. The somatic portion of the PNS controls the skeletal muscles—somatic muscles are contracted by our own volition (for this reason, another name for the somatic PNS is "voluntary"). The autonomic or involuntary portion controls muscles in the heart, digestive tract, urinary and respiratory systems, etc., which are not under our conscious control. The autonomic PNS is further distinguished form the somatic PNS in that it has pairs of neurons which interface in ganglia. One neuron originates in the spinal cord and goes to the ganglion (preganglionic). The other originates in the ganglion and has an axon that goes to a particular target. Another difference is that the autonomic PNS can also have an excitatory or stimulatory effect on its target, i.e., a signal can have one of two effects. The voluntary PNS either sends a signal causing contraction or doesn't send a signal. It sends only one type of signal that can have only one effect. The autonomic PNS can be further subdivided into the parasympathetic and sympathetic nervous systems. The sympathetic nervous system is composed of axons radiating out from the central portions of the spinal cord, while the parasympathetic nervous system has axons originating at the top and bottom of the spinal cord.

Senses and receptors

Senses obtain information about the environment through receptor cells. In some cases, such as smell, the receptor is a neuron. In others, such as taste buds, the receptors are specialized cells. The receptors act as transducers, changing energy or chemical stimulation into nerve impulses. In the eye the receptor cells are sensitive to light, and in the mouth to chemical stimulation.

Sensory processing

The information obtained through the senses is processed and evaluated in the brain. It is in the brain that a coherent picture of the world is formed from the diverse information which is constantly being transmitted to it. Only nerve impulses enter the brain. The picture that emerges is a result of integration of information—in a sense, the eye reacts to light and the ear reacts to sound, but it is the brain that "sees" and "hears."

Test Yourself

1. The vertebrate nervous system has two main parts, the _________ nervous system, which consists of the brain and spinal cord, and the ___________ nervous system, which consists of motor and sensory neurons running to other parts of the body, and the sense organs themselves.

2. The ________ nervous system ____________ information. The ____________ nervous system brings information in and links the _________ and spinal cord with the remainder of the body.

   a. peripheral, collects, nerves, brain

   b. central, integrates, peripheral, brain

   c. peripheral, integrates, central, collects

   d. central, collects, distal, brain

3. The basic cell type of the nervous system is the ___________.

   a. T-cell

   b. N-cell

   c. neuron

   d. Schwann cell

   e. dendrite

4. Of the two types of cell processes that the basic nervous system cell has, the one that conducts signals away from the cell body is the ____________. The process that conducts impulses toward the cell body is the ______________.

   a. axon, glial cell

   b. adductor, abductor

   c. neuron, glial

   d. axon, dendrite

5. The central nervous system, abbreviated as ______, consists of two main parts, the _________ and the _____________ (one or more words). In the ___________, the outer region is gray, the inner white. In the spinal cord, the _________ region is white and the _________ is gray. The white color indicates the presence of axons with a sheath of ___________.

6. In the peripheral nervous system, __________ or sensory neurons channel information from the senses to the CNS. Motor neurons, also called _________ neurons, carry information from the CNS to regions of the body (e.g. organs) where actions are carried out.

7. The motor portion of the peripheral nervous system is further subdivided into the ___________ or somatic nervous system and the ___________ or _____________ nervous system. The latter has two parts the ___________ and ______________ nervous systems, which work in opposition to each other when they both innervate a particular organ.

Questions 8–17. Identify the nervous system associated with the following functions. Possible answers: parasympathetic or sympathetic.

8. Accelerates heart rate

9. Reduces heart rate

10. Increases digestive activities

11. Reduces digestive activities

12. Causes more blood to go the heart, away from peripheral circulation

13. Empties colon and bladder

14. Increases secretion of epinephrine (adrenaline)

15. Constricts the pupils

16. Dilates (enlarges) the pupils

17. Dilates bronchioles (passages in lungs)

18. The impulses conducted by neurons are (pick all that are true)

    a. electrical currents

    b. carried along the axons and dendrites of the neuron by acetylcholine

    c. generated by ions moving across the plasma membrane of the neuron

    d. created by changes in the permeability of the plasma membrane of the neuron

19. Pick all of the following that are correct statements about neurons:

    a. The resting potential is created by a pump in the cell membrane, which moves sodium ions out of and potassium ions into the cell, and is powered by ATP.

    b. The inside of the neuron at resting potential is negatively charged compared to the outside. This is the basis of the polarization of the neuron.

    c. Depolarization occurs when the ions move across the membrane, the charge difference is lost.

    d. Depolarization and then a return to the resting potential moves progressively down the neuron—this is the nerve impulse.

20. Pick the correct statements:

    a. Schwann cells form sheaths on the neurons in the brain and elsewhere in the CNS.

    b. The spaces between Schwann cells are called nodes of Ranvier.

    c. Schwann cells form myelin sheaths on axons of neurons in the peripheral nervous system.

    d. The time required for the neuron to recover from depolarization is termed the refractory period; it accounts for the direction of movement of the impulse, since it cannot travel to membrane where depolarization has already occurred.

21. True or false: During membrane depolarization, (i) sodium ions first rush into the cell, reducing the charge difference between outside and inside to zero, (ii) the inside of the cell becomes positively charged, and (iii) finally the potassium ions, leaving the cell, return the charge inside the cell to negative compared to the outside.

22. The nodes of Ranvier

    a. allow current to jump from one node (bare spot with no Schwann cell/myelin sheath) to another, by passing through the extracellular fluid around the neuron.

    b. are swelling along nerve fibers.

    c. conduct electrical impulses only in the heart.

    d. are found only in the CNS.

23. The "jumping" of the nerve impulse from one node of Ranvier to another is called

_______________ conduction. The change in ________ causes ion channels in the plasma membrane to react.

24. The gap between one neuron and another is called a ___________. In response to membrane depolarization, ____________ are released from vesicles in one neuron and ____________________ to the other neuron.

    a. synaptic cleft, insulin, diffuse

    b. synaptic cleft, neurotransmitters, are actively transported

    c. gap junction, neurotransmitters, diffuse

    d. synaptic cleft, neurotranmitters, diffuse

25. True or false? If false, correct the statement to make it true. Neurotranmitters like acetylcholine, norepinephrine, epinephrine, and others are released from postsynaptic vesicles across the postsynaptic membrane into the synaptic cleft, where they diffuse across the gap and are taken up by receptor molecules in the presynaptic membrane.

26. Where are these neurotranmitters generally found? Identify each one as found in the CNS or peripheral nervous system (PNS).

    (i) acetylcholine

    (ii) epinephrine

    (iii) GABA

    (iv) endorphins

27. GABA stands for ____________________ (one or more words) acid, and is an _______________ CNS neurotransmitter.

    a. gibberellin, inhibitory

    b. gamma amino butyric, excitatory

    c. gamma amino butyric, inhibitory

    d. gibberellin, excitatory

28. At the junction of muscle cell and axon of a neuron ____________ is released, causing depolarization in the membrane of the ________ cell. This allows __________ ion to enter the muscle cell and the cell contracts.

29. When the neurotransmitter is released as in question 28, how is the this neurotransmitter prevented from continuing to act, causing uncontrolled contraction of the muscle cells?

30. What is the medulla oblongata? What are four functions that are controlled there? What type of functions are these?

31. Compared to other parts of the brain, which part of the brain changes the most as we trace the evolution of the brain from amphibians, to birds, to carnivore mammals such as cats and dogs, to humans?

32. The __________ of the brain is split into two halves, connected by a nerve bundle, the _____________ ___________. Each half is subdivided into four parts or lobes, the _________, ___________, ___________, and ____________ lobes.

Questions 33–41. Matching.

Possible answers: rhombencephalon, nuclei, mesencephalon, prosencephalon, thalamus, ganglia, sensory, motor, cerebellum.

33. ascending tracts in spinal cord

34. descending tracts in spinal cord

35. midbrain

36. hindbrain

37. forebrain

38. clumped cell bodies in CNS

39. clumped cell bodies in PNS

40. center for coordination of movement, balance

41. processes sensory information in brain

Questions 42–48. Matching.

Possible answers: frontal lobe, parietal lobe, occipital lobe, temporal lobe, pons, prefrontal cortex, hypothalamus, thalamus, limbic system, right hemisphere.

42. integrates body activities such as pituitary secretions, temperature and respiration, heartbeat—in short, responsible for homeostasis.
43. location of visual cortex
44. location of auditory cortex
45. location of somatosensory cortex
46. higher mental activities and some sensory integration
47. hemisphere associated with left side of the body
48. lobe associated with speech (Broca's area)
49. Information received by the senses is interpreted in a way unique to that sense—for example ____________ on the eyeball is perceived as light.
50. List several types of (i) internal and (ii) external information that is sensed.
51. Motion and balance involve the ____________ ____________ (two words) in the ear. Movement of fluid in these chambers is detected by special sensory ________.
52. Taste and smell involve ____________ ____________ in the mucous membranes. In the case of ________, the receptors are actually neurons, whereas in ________ the receptors are specialized cells found in _______ buds.
53. Briefly describe the process of hearing a sound.
54. How are insect eyes different from human eyes? Name an invertebrate animal which has an eye similar to that of humans.
55. What does the eye detect? The ear? What are the rods and cones in the eye used for?
56. What are the two forms of visual pigment in vertebrates?
57. How is light from the two eyes transmitted to the brain?
58. What is a lateral line system?
59. What is the eustachian tube? What does it do?
60. How does the human eye focus?

Check Yourself

1. central, peripheral. **(General)**

2. b. central, integrates, peripheral, brain. Always connect the central nervous system with words like "integrate." The central nervous system (CNS) consists of the brain and the spinal cord. Here information is received from the peripheral nervous system and integrated. The peripheral nervous system (PNS) is the way the organism reaches out to gather information about its internal and external environment **(General)**

3. c. neuron **(General)**

4. d. axon, dendrite. Students sometimes fail to keep axons and dendrites straight. Here's something that might help. Dendrite comes from the Greek word for trees—trees have roots, roots conduct water and nutrients toward and into the tree, hence it makes sense that dendrite also direct something (impulses) toward and into the neuron. They are, in a sense, the "roots" of the neuron. **(General)**

5. CNS, brain, spinal cord, brain, outer, inner, myelin. Myelin is a fatty protein that sheaths nerve fibers. It acts as an insulator, something like the plastic coating on an electrical cord. Myelin sheaths are extensions of Schwann or glial cells. **(General)**

6. afferent, efferent. The "a-" is a clue that the word refers to something going toward an object; conversely, the "e-" is a clue that the word refers to something moving away from an object. **(Peripheral nervous system)**

7. voluntary, autonomic, involuntary, sympathetic, parasympathetic. It should make sense that these are parts of the motor portion of the peripheral nervous system—all these nerves cause actions to be carried out. **(Peripheral nervous system)**

8. sympathetic **(Peripheral nervous system)**

9. parasympathetic **(Peripheral nervous system)**

10. parasympathetic (increases motility and secretion) **(Peripheral nervous system)**

11. sympathetic **(Peripheral nervous system)**

12. sympathetic **(Peripheral nervous system)**

13. parasympathetic **(Peripheral nervous system)**

14. sympathetic **(Peripheral nervous system)**

15. parasympathetic **(Peripheral nervous system)**

16. sympathetic **(Peripheral nervous system)**

17. sympathetic

    The sympathetic helps the body prepare for action—the "flight-or-fight" reaction—by increasing heartbeat and respiration, increasing the size of the lungs passageways (dilation of bronchioles or bronchodilation) to help breathing, shunting blood to the heart and away for the digestive tract and skin. Less vital functions such as digestion are curtailed. This means inhibiting the secretions and motility of the digestive tract. Secreting epinephrine (adrenaline) helps accomplish some of these tasks, so secretion of that hormone is also stimulated. By contrast, the parasympathetic nervous system tends to slow things down, conserving energy and stimulating day-to-day life functions such as digestion. The sympathetic nervous system is useful if a bear is hunting you; the parasympathetic, if you have hunted and are eating a bear.

When both nervous systems enervate an organ, they usually have antagonistic effects, such as are listed below:

<u>Sympathetic versus parasympathetic effects</u>

sympathetic	parasympathetic
dilates pupil	contracts pupil
bronchodilation	bronchoconstriction
speeds heart rate	slows heart rate
relaxes urinary bladder	contracts urinary bladder

There are exceptions to this generalization: e.g., both systems stimulate salivation, although to different degrees. **(Peripheral nervous system)**

18. c, d. Nerve impulses are not electrical currents like those flowing the household electrical wiring. Nerve impulse result from ions moving across the plasma membrane, changing the difference in ion concentration (and thus charge) between the inside and the outside of the neuron. **(Nerve impulses)**

19. All of these statements except (a) are correct. Nerve impulses travel toward the central bodies of neurons along dendrites and away along axons. The impulse itself is electrochemical in nature, and is slower than and distinctively different from the flow of electricity (electrons) seen in household wiring. A potential (voltage) difference exists between the inside and outside of the plasma membrane of a neuron. The electrochemical gradient that establishes this difference is due to the pumping of sodium ions out of the neuron and potassium ions into the neuron. The potassium ions flow freely back out across the membrane, but the sodium ions cannot directly diffuse across the membrane, leaving a net negative charge inside the cell. **(Nerve impulses)**

20. b, c, d. **(Nerve impulses)**

21. True. When a neuron "fires" after being stimulated, an action potential is generated which depolarizes the membrane, opening entry points for sodium ions. The sodium ions flow into the cell, and the voltage difference between the inside and outside of the cell disappears. These gates open sequentially along the neuron, propagating the impulse. As some time is require to regenerate the potential (charge) difference by transferring ions across the plasma membrane, the impulse can only move forward to regions where the action potential has not yet had an effect. The membrane regions just crossed by the impulse are unresponsive. This period of non-responsiveness is termed the "refractory." **(Nerve impulses)**

22. a. are swelling along nerve fibers **(Nerve impulses)**

23. saltatory (from "saltation," or jumping), voltage (or potential difference) **(Nerve impulses)**

24. d. synaptic cleft, neurotransmitters, diffuse. At the end of the axon the action potential causes chemicals (neurotransmitters) to be released from membrane-bound vesicles in the axon. The chemicals diffuse across the gap (synaptic gap) to excite another neuron, or a muscle cell. Depending on the circumstances,

the action can be inhibitory or excitatory. The former works against depolarization of the membrane, while the excitatory action depolarizes the membrane. Mixed signals can be received, and the sum total of their effects determines the integrated response of the neuron or cell. **(Neurotransmitters)**

25. The statement is false since it uses postsynaptic where presynaptic should be used, and visa versa. The corrected statement reads (with corrections underlined):

    Neurotranmitters like acetylcholine, norepinephrine, epinephrine and others are released from presynaptic vesicles across the presynaptic membrane into the synaptic cleft, where they diffuse across the gap and are taken up by receptor molecules in the postsynaptic membrane. **(Neurotransmitters)**

26. (i) PNS (as in motor neurons)

    (ii) CNS

    (iii) CNS

    (iv) CNS

    Some are found in the PNS, others are found only in the CNS. Acetylcholine is responsible for stimulating muscle cells, and is secreted by motor neurons. Epinephrine, norepinephrine and serotonin are mainly found in the CNS. GABA (gamma aminobutyric acid) is found only in the CNS, as are neuropeptides called endorphins. **(Neurotransmitters)**

27. c. gamma amino butyric, inhibitory **(Neurotransmitters)**

28. acetylcholine, muscle, calcium **(Motor responses)**

29. Acetylcholinesterase (note the -ase ending) degrades the acetylcholine molecules shortly after they are released, assuring that only freshly released acetylcholine is actively causing muscle membrane depolarization. **(Motor responses)**

30. The medulla oblongata is really a projection of the spinal cord into the skull; it is part of the hindbrain. In this region, breathing, heartbeat, digestion and vomiting are controlled; these are all autonomic functions. **(Brain)**

31. The cerebral cortex has shown the greatest number of changes through the course of evolution. **(Brain)**

32. cerebral cortex, corpus callosum, frontal, temporal, occipital, parietal **(Brain)**

33. sensory (brings information into CNS) **(Central nervous system)**

34. motor (takes information away from CNS) **(Central nervous system)**

35. mesencephalon (mes- meaning middle, and -encephalon for brain) **(Brain)**

36. rhombencephalon **(Brain)**

37. prosencephalon **(Brain)**

38. nuclei **(Central nervous system)**

39. ganglia **(Peripheral nervous system)**

40. cerebellum **(Brain)**

41. thalamus **(Brain)**

42. hypothalamus. Vertebrate brains for the past 500 million years have characteristically had three parts: the hindbrain, midbrain, and forebrain. (These are more technically termed the rhombencephalon, mesencephalon, and prosencephalon, respectively.) The brains of fishes consist mainly of hindbrain (pons, medulla oblongata, and cerebellum). These may be regarded as part of the spinal cord that intrude into the brain case. Fish have evolved other parts of the brain for vision and smell. The forebrain becomes larger in the amphibians and reptiles. In these animals as well as birds and mammals, the forebrain consists of the diencephalon (containing the hypothalamus and thalamus) and the telencephalon (containing the cerebrum). The development of the cerebrum is associated with the acquisition of higher brain functions, including reasoning. It is a dominant anatomical feature of mammalian brains, and is responsible for much of the size difference seen in comparing mammalian brains with those of other vertebrates. **(Brain)**

43. occipital **(Vision and the eye)**

44. temporal. **(Hearing and the ear)**

45. parietal (somatosensory cortex receives input from touch and pressure receptors) **(Sensory processing)**

46. prefrontal cortex **(Brain)**

47. right (and the left hemisphere is associated with activities on the right side of the body) **(Brain)**

48. frontal **(Brain)**

49. pressure **(Senses and receptors)**

50. (i) blood pressure, carbon dioxide level in the blood (carotid receptors), body temperature, muscular movement (stretch receptors associated with muscles), position of limbs; (ii) vision, hearing, touch, taste, smell. **(Senses and receptors)**

51. semicircular canals, cilia (specifically kinocilia and stereocilia) **(Hearing and the ear)**

52. chemical receptors, smell, taste, taste **(Senses and receptors)**

53. A sound is created by air being formed into waves which take the form of alternating regions of greater and lesser compression (density) of air. When a book falls from the desk to the floor, mechanical vibrations compress air molecules. A pattern is set up, areas of slightly compressed (denser) moving air molecules alternate with less compressed regions of air. These make up a repeating pattern called sound waves. The frequency of the sound is determined by the wave pattern. These waves enter the ear, where they strike the eardrum, cause it to vibrate, and in turn communicate the vibrations to a series of three bones which act as tiny levers amplifying the sound: the hammer (malleus, as in mallet), the anvil (incus), and the stirrup (stapes—easy to recall because a stirrup resembles a paper staple). The stirrup vibrates a membrane, the oval window, which forms part of the wall of the cochlea. Receptor hair cells, called cilia, on the basilar membrane are found under the tectorial membrane, lining the cochlea, and they pick up the sound energy being transmitted to the endolymph fluid. As they bend, they depolarize associated neurons, and send an impulse we interpret in our brain as sound. **(Hearing and the ear)**

54. Other types of eyes are found throughout the Kingdom Animalia. Insects have compound eyes, with numerous lenses like the facets on a diamond. Other eyes are simple, and do little more than detect levels of light. Among invertebrates, the eye of the octopus is strikingly similar to that of vertebrates, although evidence suggests eyes in the octopus and the vertebrate evolved independently. **(Vision and the eye)**

55. The eye detects photons, that is, particles of light of different wavelengths (energies). The ear detects sound waves. The eye contains rods and cones. In the eye rods are more sensitive to light but cannot distinguish color. The cones are less light sensitive, but there are cones sensitive to each of the colors red, blue, and green. The differences in light sensitivity between rods and cones relate to differences in the type of pigment present.

    The eye of the vertebrates is layered. Each layer has a distinctive role. The outer layer of connective tissue (sclera) holds the eye together and is the transparent layer over the front of the eyeball (cornea). The middle layer (choroid) contains blood vessels, and forms the iris, the colored part of the eye with the pupil (opening) in the center. A lens in the eye, controlled by small muscles, allows light to be focused on the retina, a layer inside the eyeball. The retina has rods and cones, containing different visual pigments, and is sensitive to low light levels (rods) and to color (cones). The pigments change on absorption of light, affecting the shape of the retina. This in turn starts an action potential, and generates a nerve impulse along the optic nerve. The impulse is interpreted in the brain (visual cortex) and processed to form an image. (**Senses and receptors**)

56. The visual pigments are a derivative of carotene (a plant pigment), rhodopsin, and iodopsin (photopsin). The pigments are chemically and structurally altered by interaction with light. (**Vision and the eye**)

57. Light is seen by the brain in the form of nerve impulses produced by light (photons) striking the pigments in the retina. The impulses are conducted by the optic nerves. Information from the two eyes is mixed to a degree due to the redundant nature of the optic nerve plexus. Instead of the information from the right eye only going to the left brain and vice versa as two totally distinct and separate images, some of the image from the left eye is combined with the main right eye image, and some of the right eye image is combined with the main left eye image. (**Vision and the eye**)

58. Some aquatic amphibians and many fish have lateral lines. Lateral lines act as a sensory system on both sides of the body that can detect changes in pressure (which includes sounds) underwater. Hair cells (neuromasts) are the specialized sensors for this system. (**Senses and receptors**)

59. The eustachian tube vents the side of the eardrum facing the center of the head, allowing atmospheric pressure to become equal on both sides. Unless a person has a cold or allergy that cause swelling and blockage the pressure is equalized. In the case of illness the tube is swollen shut, pressure builds, and hearing is less acute since the eardrum's movement is dampened by unequal pressure on either side. (**Hearing and the ear**)

60. Small muscles attached to the lens of the eye pull or relax, changing the shape and thus the focus of the lens. (**Vision and the eye**)

Grade Yourself

Circle the question numbers that you had incorrect. Then indicate the number of questions you missed. If you answered more than three questions incorrectly, you will then have to focus on that topic. (If a topic has less than three questions and you had at least one wrong, we suggest you study that topic also. Read your textbook, a review book, or ask your teacher for help.)

Subject: Nervous System and Senses

Topic	Question Numbers	Number Incorrect
General	1, 2, 3, 4, 5	
Peripheral nervous system	6, 7, 8, 9, 10, 11, 12, 13, 14, 15, 16, 17, 39	
Nerve impulses	18, 19, 20, 21, 22, 23	
Neurotransmitters	24, 25, 26, 27	
Motor responses	28, 29	
Brain	30, 31, 32, 35, 36, 37, 40, 41, 42, 46, 47, 48	
Central nervous system	33, 34, 38	
Vision and the eye	43, 54, 56, 57, 60	
Hearing and the ear	44, 51, 53, 59	
Sensory processing	45	
Senses and receptors	49, 50, 52, 55, 58	

Water Balance; Temperature Regulation; Musculoskeletal System

Brief Yourself

Osmoregulation

Osmoregulation is the process by which animals regulate the gain and loss of water to main homeostasis. Animals are composed of more than 70 percent water, contains many dissolved solutes; the concentration of solute of each side of the plasma membrane determines the direction of the net movement of water. (Recall the topics of diffusion and osmosis.) Animals and animal cells usually attempt to compensate for this movement of water, and keep water and solute concentrations within physiological ranges. In most animals, not only is the osmolarity of the cell interior regulated, but also the interior of the body, which provides fluid in which cells are bathed (interstitial or extracellular fluid, circulatory system).

Excretion

Excretion is the process by which animals get rid of excess water and nitrogenous (nitrogen-containing) wastes produced during metabolism. (The principal source of this nitrogen is protein.) The centerpiece of the vertebrate excretory system is the kidney, and the kidney is organized around the nephron (consisting of a tubule and blood vessels). This system also includes the large and small intestines, the urinary bladder, and associated ureters.

Thermoregulation

Animals live in an environment in which temperature varies, and they have evolved many responses to this variation. Body temperature is important, since it affects the rates of chemical reactions and the effectiveness of enzymes. Recall that heat is due to the motion of molecules—the higher the temperature, the faster the molecules move. Animals gain or lose heat to their environment in four ways: conduction (direct transfer of heat, as from an ice pack), radiation (transmission of heat by electromagnetic waves, such as infrared light, like sitting in the sunlight or before a fire), convection (a fluid like water or air carries heat, and moves past an animal—a fan blowing air), and evaporation (water turns from liquid to vapor, and in making this phase change, carries away heat—sweating). Ectotherms (poikilotherm) have a body temperature related to the temperature of their environment, and rely on behavior to control it. Generally, ectotherms have a lower body temperature and a correspondingly lower metabolic rate than endotherms.

Musculoskeletal system

The musculoskeletal system of vertebrates consists of bones, joints, connective tissue such as ligaments and tendons and muscles. The body of mammals and some other vertebrates can be divided into the appendicular (limb) skeleton, and the axial (central body) skeleton.

Muscle contraction

Muscle makes up more of the body weight of a vertebrate than any other single tissue. Muscle fibers (each one a very long, single cell) make up each skeletal muscle, looking in cross section like the ends of uncooked spaghetti. Each fiber is bound by a special cell membrane (sarcolemma), which can function like the membrane of a neuron and carry an action potential.

Each cell in turn contains several thousand myofibrils, long filaments that run the length of the muscle cell. A form of endoplasmic reticulum called sarcoplasmic reticulum ensheathes each myofibril.

Test Yourself

1. Animals that maintain internal osmotic conditions different from that of the surrounding environment are termed ______________. Animals which allow their internal environment to follow the osmotic conditions of the external environment are called ___________.

2. Freshwater fishes are ___________ (hypotonic, isotonic or hypertonic) relative to their environment and tend to __________ (lose, gain) water.

3. Marine fishes are ___________ (hypotonic, isotonic or hypertonic) relative to their environment and tend to __________ (lose, gain) water.

4. Excretion involves removing ____________wastes and other materials from the body, while retaining needed __________ and salts.

5. The nitrogen in wastes is derived mainly from _________ in the diet.

6. Fishes excrete most of their nitrogenous wastes as _________; sharks, some other marine fishes, amphibians, and mammals excrete mainly __________; insects, birds, and most reptiles excrete __________.

   Possible answers: urea, carbon dioxide, ammonia, uric acid, feces.

7. The basic unit of the vertebrate kidney is the ___________.

8. Trace the passage of waste products by listing the following in the proper order:

   a. loop of Henle

   b. collecting duct

   c. distal convoluted tubule

   d. Bowman's capsule

   e. proximal convoluted tubule

   f. glomerulus

9. As __________ ions are actively pumped out of the tubule, _________ ions passively follow, as does _______ by osmosis.

10. The permeability of the collecting duct to water is controlled by the ____________ hormone.

11. The kidney may be grossly divided into two regions, the outer ________ and the inner _________. In which region is the concentration of solutes highest?

12. Insects excrete _________ ions into the ____________ tubule, causing water and waste products to follow. Reabsorption of water and ions occurs in the ________, part of the digestive tract. (Possible answers: solute, water, potassium, sodium, chloride)

13. All but the simplest animals have an internal environment in which cells are bathed in _____________ fluid.

14. Mammals and birds have highly developed kidneys and can excrete urine that is ________ (hypotonic, isotonic, hypertonic) compared to body fluids.

15. The underlying means by which the vertebrate kidney works is that the tubule moves through regions of varying osmotic conditions—the regions of higher solute concentrations cause water to leave the tubule. The high concentration of this material is provided by active pumping of __________ (sodium, chloride, urea) ions from the tubule, and _________ (sodium, chloride, urea) from the collecting duct.

16. Water that is derived from using food in the body is called __________ water.

17. Sharks become _________ (hypotonic, isotonic, hypertonic) with respect to their environment by building up a high concentration of ________ (ammonia, urea, uric acid) in their body tissues.

18. Heat can be transmitted in fours ways. Name them, and give an example of each from everyday life.

19. Identify the following as ectothermic or endothermic animals: earthworm, frog, turtle, sponge, sea star, dog, human, robin, cow.

20. Can ectothermic animals be correctly termed "cold-blooded"? Why or why not?

21. True or false: Hibernation is a state in which body temperature simply falls.

22. Name some animals which hibernate. Do bears hibernate?

23. List three examples of adaptations in animals to severe environments.

24. A kangaroo rat may go all of its life without drinking a drop of water. How does the animal survive?

25. As blood leaves the foot of a duck adapted to cold weather, and heads toward the body, it passes closely by outgoing blood in an adjacent vessel. Heat is transferred from the warm outgoing blood to the cold incoming blood. The is an example of __________________ heat exchange.

26. If a person were stranded on a life raft without fresh drinking water, would it be a good idea to drink salt water to survive? Why or why not?

27. How are bones joined to muscles?

28. How are bones linked to bones?

29. Blood cells are made in the ____________ of long and other bones.

30. Name the two types of filaments found in skeletal muscle.

31. A muscle fiber is actually a single _________, often with several nuclei.

32. Each muscle cell contain thousands of ____________, surround by _____________ reticulum.

33. Each myofibril is composed of _________________ (means "muscle unit"), which contain the two type s of filaments.

34. Which of the following are needed for a muscle to contract:

    glucose, ATP, ADP, calcium ions, signal from the nervous system to the motor unit.

35. When a skeletal muscle contracts the distance between the ____ -lines shortens as the thick and thin ___________ slide over each other.

36. The bridge between a thick and thin filament consists of a head of the protein ________ with an ________ molecule attached.

37. What bands can be seen in a muscle under the electron microscope?

38. Which band has only thin filaments? Which band has only thick filaments?

39. What are the two proteins involved with calcium ions in the control of muscle contraction?

    a. troponin

    b. trophonin

    c. myosin

    d. tropomyosin

40. True or false; if false, explain why. The force of a muscles contraction is due to how strongly each muscle cell is stimulated by the nervous system.

41. A neuron and the muscle cells it innervates is called a _______ _______ (two words).

42. Muscle cells would likely contain many of what subcellular organelle?

43. True or false: Exercise will strengthen a muscle, and make it able to push harder.

44. Bone tissue is built up by cells known as ___________ and broken down by __________.

45. What are some problems land animals face in maintaining water balance?

46. Discuss endothermic animals and control of body temperature.

Check Yourself

1. osmoregulators, osmoconformers. Animals may adjust to environmental demands in several ways. Osmoconformers, such as many marine invertebrates and sharks among the vertebrates, have an osmolarity within their cells that is about the same as the osmolarity of their environment. By contrast, osmoregulators maintain their internal osmolarity even if the osmolarity of their environment changes. These animals are not isosmotic with their environment (i.e., different concentrations of solutes exist in the animal and in the environment). Osmoregulators include some marine animals, all freshwater animals, and most terrestrial animals. **(Osmoregulation)**

2. hypertonic, gain. Different environments present different challenges to animals. A fish in a fresh water environment is hypertonic compared with the water around it. It gains water into its body as water moves across its gills; consequently, freshwater fish never drink water. Some solutes are lost through the skin, and solutes are gained from food. Some needed solutes are taken in by specialized absorption cells in the gills. The problem for a freshwater fish is to get rid of excess water, which it does by excreting a large volume of urine. **(Osmoregulation)**

3. hypotonic, lose. A saltwater fish has the opposite problem. It is hypotonic to its environment. Saltwater moving across its gills osmotically removes water from the fish's body; this loss is countered by drinking sea water. The sea water, however, carries with it an excess of solutes, which are excreted in concentrated urine and from the gills where specialized cells excrete excess solute such as sodium ions. **(Osmoregulation)**

4. nitrogenous (nitrogen-containing), water. Excretion is how animals get rid of excess water and nitrogenous (nitrogen-containing) wastes produced during metabolism. The principal source of this nitrogen is protein. Wastes may occur as ammonia, uric acid or urea. It is a closely controlled process, since it is important that some substances, such as ions, are conserved and retained in the body. Excretion is carried out by a varicty of mcans, ranging from the contractile vacuole in the protist *Paramecium* to the kidneys of mammals. **(Excretion)**

5. protein **(Excretion)**

6. ammonia, urea, uric acid. Fishes excrete most of their nitrogenous wastes as ammonia; sharks, some other marine fishes, amphibians, and mammals excrete mainly urea; insects, birds, and most reptiles excrete uric acid. **(Excretion)**

7. nephron **(Vertebrate excretory systems)**

8. f, d, e, a, c, b. Recall the meaning of the words proximal (near, as in approximate) and distal (as in distance) to help you keep these regions straight. The kidney has an outer region (cortex) and inner region (medulla). These regions are grossly visible and distinct to the eye. Renal arteries carry blood to the kidneys. There, blood flows through a knot of capillaries called the glomerulus. Pushed by blood pressure, plasma-like fluid from the blood is forced into Bowman's capsule, which surrounds the glomerulus. (The relationship of the glomerulus to the capsule can be envisioned by forming one hand into a fist and surrounding it with the opened palm of the other hand; the fist is the glomerulus, the opened hand, Bowman's capsule. Two hands together—glomerulus and capsule—are called the Malpighian corpuscle.) Fluid then moves through a tubule that can be divided into several functional regions: proximal convoluted tubule, loop of Henle, distal convoluted tubule, and the collecting duct. The tubule dips down from the cortex of the kidney into the medulla, and then comes back up into the cortex, passing through zones of different osmotic conditions, which contribute to the reabsorption of materials from the fluid.

Inner regions of the kidney have more solutes and thus cause water to move from the tubule. This allows the kidney to make hypertonic urine. **(Vertebrate excretory systems)**

9. sodium, chloride, water. **(Vertebrate excretory systems)**

10. antidiuretic. Diuresis is the passing of urine, hence antidiuretic means reducing the production or volume of urine. As the fluid moves through these regions, water and many dissolved substances such as glucose and ions of various salts are reabsorbed by the surrounding network of capillaries. Fluid in the proximal convoluted tubule loses sodium by active transport, with chloride ions and water following passively by ionic attraction and osmosis, respectively. As the loop descends into the medulla, water is lost into the surrounding zone of high solute concentration. (The solutes are sodium and chloride actively pumped out the loop of Henle, and urea diffusing out of the neighboring collecting duct.) The ascending portion of the loop of Henle, which leads to the distal convoluted tubule, is closed to water movement, so as sodium and chloride ions are actively pumped out, water cannot passively follow. In the distal convoluted tubule, more water is lost from the tubule into the surrounding zone (and thus back into the bloodstream), and this loss continues in the collecting duct. Antidiuretic hormone (ADH) can act at this point. ADH increases the permeability of the collecting duct wall to water. Without ADH, more water is retained within the tubule, and more dilute urine excreted. **(Vertebrate excretory systems)**

11. cortex, medulla, medulla. **(Vertebrate excretory systems)**

12. potassium, Malpighian, hindgut. Invertebrates can have one of several specialized excretory systems. Flatworms such as planaria have protonephridia and segmented worms have metanephridia, both of which function to conserve salts, proteins, and sugars while getting rid of excess materials. Insects have the more sophisticated Malpighian tubules. Potassium ions are actively pumped from the hemolymph into the tubule, and are followed by wastes and water. The tubules drain into the hindgut of the digestive system, where water and required ions (salts) are reabsorbed. **(Invertebrate excretory systems)**

13. interstitial or extracellular **(Osmoregulation)**

14. hypertonic **(Vertebrate excretory systems)**

15. sodium, urea **(Vertebrate excretory systems)**

16. metabolic **(Osmoregulation)**

17. isotonic, urea **(Vertebrate excretory systems)**

18. Conduction: sitting on a cold (or hot) floor.

    Convection: being outside on a windy day (windchill).

    Radiation: warming yourself in front of a fireplace or by lying in the sunlight.

    Evaporation: feeling chilled while wearing wet clothes. **(Thermoregulation)**

19. The first five are ectotherms, the last four are endotherms. **(Thermoregulation)**

20. While an ectotherm may seem cold-blooded when handled by an endothermic animal (such as a human), the term "cold-blooded" is misleading. It suggests the animals body temperature is always cold (picking up a basking snake will suggest otherwise) and that it slavishly follows the temperature of the environment. Ectothermic body temperature may be strikingly modified by behavior such as vigorous activity. Bumble bees and some other insects exercise their wing muscles before flight to raise their body temperature to a point where the muscles can operate efficiently. Honey bees huddle in groups to conserve

heat, and can use their wings as fans to cool a hive in hot weather. Reptiles, such as snakes, lizards, and many turtles, bask in the sun to raise body temperature. (If a snake cannot find a warm place after a meal, the food in its stomach may actually begin to decay before it is digested.) Fishes such as tuna have heat conservation mechanisms that allow them to conserve the heat produced by muscular contraction. **(Ectotherms)**

21. False. This is an incomplete description of hibernation, in which the animal experiences a controlled drop in body temperature. The body temperature is still regulated, just at a lower level. Torpor refers to a state in which metabolism is slowed, and occurs in several forms. In hibernation, the body temperature falls and is regulated at a lower temperature. This conserves food reserves such as body fat and allows animals to wait out a period of unfavorable weather, such as winter. Many rodents hibernate, but bears do not. Bears reduce their body temperature by only a few degrees, and their metabolic rate does not drop as much as it does in hibernating rodents. Related to hibernation is a state called diurnation, during which there is a short-term drop in body temperature, part of a daily cycle. This is seen in animals such as bats. **(Torpor)**

22. Rodents such as chipmunks, squirrels, and some mice hibernate, as do hamsters. Bear do not truly hibernate, since their body temperature and metabolic rate decrease less than in hibernating animals. (Polar bears stay active all winter, by the way.) **(Torpor)**

23. (i) countercurrent heat exchange, (ii) thick fur or layer of subcutaneous fat, (iii) ability to conserve water by producing very concentrated urine. Other correct answers are possible. Under extreme environmental conditions, some ectotherms retreat to protected areas such as burrowing in the mud, as some frogs do during summer estivation. Many animals have special adaptations for extreme environments such as the arctic or the deserts. Some endotherms have adaptations to help conserve body heat, such as air-trapping fur or feather, and a layer of insulating fat beneath the skin. The flow of blood to the skin and extremities can also be modified to increase (vasodilation) or decrease (vasoconstriction) the loss of heat. Some endotherms (both birds and mammals) have a countercurrent heat exchange system for their limbs, where warmer blood flowing out warms incoming colder blood in nearby vessels. This reduces the amount of body heat leaving the limb, part of which would be lost to the environment. Some animals have additional adaptations such as antifreeze compounds in their bodies (some frogs, arctic fishes, and some insects). **(Cold and hot environments)**

24. The kangaroo rat conserves water by remaining in its burrow during the hot part of the day, and by producing very dry feces and concentrated urine. It is thus able to survive on water found in food and produced by the metabolism of food (especially food rich in fat). **(Cold and hot environments)**

25. countercurrent **(Cold and hot environments)**

26. Drinking salt water would hasten the person's death from dehydration. The amount of water in salt water is not enough for the kidneys to get rid of the salts which accompany it. A person would end up with more salts in the body, and become thirstier than before drinking the salt water. **(Vertebrate excretory systems)**

27. Bones and muscles are joined by tendons. **(Musculoskeletal system)**

28. Bones are joined to other bones (as at joints) by ligaments. **(Musculoskeletal system)**

29. marrow **(Musculoskeletal system)**

30. thin and thick filaments **(Muscle contraction)**

31. cell **(Cold and hot environments)**

32. myofibrils, sarcoplasmic. A single myofibril is found to be composed of sarcomeres (literally "muscle units"), with two types of filaments, thick and thin. In a relaxed muscle, minimum overlap occurs between the two types of filaments. When the muscle contracts, the filaments overlap, shortening the muscle. (Remember a muscle can only do work by contracting and pulling—it cannot expand and "push" anything.) Contraction occurs as follows. A protein myosin head on a thick filament has an attached ATP molecule. It forms a bridge with the actin molecule of an actin (thin) filament. As a phosphate group is released from the ATP and ADP is formed, the bridge bends, moving the thin filament and shortening the sarcomere. Another molecule of ATP frees the myosin from its connection with the actin filament. This second ATP molecule also prepares the myosin molecule for another contraction. **(Muscle contraction)**

33. sarcomeres **(Muscle contraction)**

34. All but glucose (energy is supplied by ATP) and ADP are needed for muscle contraction. The sarcomere is made up a series of bands, the I and A bands, as well as the H zone and Z lines, and a line in the middle termed "M." These regions can be seen in an electron microphotograph of skeletal muscle. Thin filaments are anchored on the Z-line, the I band has only thin filaments and the A band has only thick filaments. Contraction pulls the Z-lines closer together, reflecting the sliding of thin over thick filaments. **(Muscle contraction)**

35. Z, filaments **(Muscle contraction)**

36. myosin, ATP **(Muscle contraction)**

37. I and A bands **(Muscle contraction)**

38. The A band is made up of thick filaments, the I band of thin filaments. **(Muscle contraction)**

39. a. troponin and d. tropomyosin. Contraction is regulated by calcium ions and two proteins, troponin and tropomyosin. In the absence of calcium ions, the tropomyosin molecules block the bridging locations on the actin (thin filament) molecules. The binding of calcium ions to troponin causes a shift in the blocking tropomyosin molecules, allowing the myosin head of a thick filament access to the bridging site. Calcium ions are released to perform this function at the command of a signal from the nervous system, delivered to the neuromuscular junction of the muscle cell. The axon of a motor neuron and the muscle fibers under that neuron's control form a motor unit. The force of a muscle's contraction depends on how many motor units are stimulated to contract. **(Muscle contraction)**

40. False. A muscle fiber is stimulated to contract; it is not a matter of degree. The cell either contracts or it does not. The strength of the muscle's response is determined by the number of muscle cells stimulated to contract. **(Muscle contraction)**

41. motor unit **(Muscle contraction)**

42. Muscle cells are likely to be rich in mitochondria, since the cells use a lot of energy. **(Muscle contraction)**

43. False. The muscle will indeed strengthen with exercise, but it will never push—muscle cells can only contract (shorten), not expand, to do work. **(Muscle contraction)**

44. Osteoblasts, osteoclasts. Bone, another component of the musculoskeletal system, is found in bony fishes (class Osteichthyes) and other vertebrates. Bones are made up of calcium and phosphorus salts, and may contain marrow for the production of blood cells. Bone is laid down by cells called osteoblasts, and taken up by osteoclasts. Cartilage, an elastic type of connective tissue, is found as part of the skeletal system in

several forms, and in places such as the pinna of the ear and the tips of the nose and sternum. **(Musculoskeletal system)**

45. Land animals face the problem of obtaining and conserving water, while getting rid of waste products. Water can be obtained from the diet, from drinking, or as a byproduct of oxidation of foods in the mitochondria (metabolic water). Excess solutes (from the diet) must be eliminated and essential solutes conserved and not lost excessively in the urine or sweat (in animals that perspire). **(Osmoregulation)**

46. Endotherms (homeotherms), which consist of the mammals and birds, keep a fairly constant internal temperature. (Normal human body temperature, for example, varies from about 97 to 99 degrees F, it is not a constant 98.6 degrees F.) Heat is mainly derived from the metabolism of food. Control of body temperature is managed by the hypothalamus. Skin receptors provide information about the environment, and the hypothalamus itself senses the blood temperature. By comparing temperature information, the hypothalamus can initiate responses to cool (e.g., sweating in humans, panting in dogs) or warm (e.g., shivering, increase in the metabolic rate) the body as needed. **(Thermoregulation)**

Grade Yourself

Circle the question numbers that you had incorrect. Then indicate the number of questions you missed. If you answered more than three questions incorrectly, you will then have to focus on that topic. (If a topic has less than three questions and you had at least one wrong, we suggest you study that topic also. Read your textbook, a review book, or ask your teacher for help.)

Subject: Water Balance; Temperature Regulation; Musculoskeletal System

Topic	Question Numbers	Number Incorrect
Osmoregulation	1, 2, 3, 13, 16, 45	
Excretion	4, 5, 6	
Vertebrate excretory system	7, 8, 9, 10, 11, 14, 15, 17, 26	
Invertebrate excretory system	12	
Thermoregulation	18, 19, 46	
Ectotherms	20	
Torpor	21, 22	
Cold and hot environments	23, 24, 25, 31	
Musculoskeletal system	27, 28, 29, 20, 44	
Muscle contraction	30, 32, 33, 34, 35, 36, 37, 38, 39, 40, 41, 42, 43	

Animal and Human Reproduction and Development

Brief Yourself

Reproduction can be sexual or asexual. While the former offers the advantage of greater genetic variability in the offspring, it incurs the costs of finding a mate; in fact, it can be argued that asexual reproduction is more likely to be profitable from the female's point of view. Nonetheless, while asexual reproduction is not uncommon, and is found in invertebrates and even some vertebrates, the sexual form retains predominance.

Gametogenesis is the production of sperm and eggs, for many animals the sole haploid links in their life cycle. Sperm are equipped to compete with other sperm from that particular male as well as with sperm from other males. Eggs provide the initial nutrition for the developing embryo. Yolk stores vary greatly in size, from minimal to ponderous (such as in a ostrich egg). Fertilization is the nuclear union of two gametes to form a genetically unique zygote, and may be accomplished externally (as in many fish) or internally (as in humans).

The embryo is a very dynamic organism. Development is a matter of forming different tissues in different patterns. Tissues roll over each other, fold, separate, join, and are discarded (programmed cell death) in the developing embryo. Development must work within the constraints of evolutionary history; that is, patterns of development evolve out of the patterns of ancestors and are not made from scratch in the most efficient manner. The exact mode isn't as important as the fact that it works, producing a viable offspring.

In development, cells differentiate to form specialized cells, tissues and organs. Morphogenesis results in the formation of an adult body plan, and the organism grows in size as well.

Reproduction in mammals is a complex topic, as hormones, reactions to external cues, and social behavior are all involved. While humans share this complex biology, human reproductive issues are further complicated by religious and ethical issues, as well as the complexity of society and the human family.

Test Yourself

1. Aphids can reproduce _________ or ___________. They reproduce ___________ when numbers are relatively low with an ample supply of food, but resort to ______ reproduction when aphid numbers increase and the competition for food also increases.

2. Sexual reproduction results in a greater __________ ___________ (two words) among offspring, while asexual reproduction produces copies of the parent.

3. Asexual reproduction is confined to lower animals. True or False?

4. Define and give an example of :

   (i) Oviparity (oviparous animals)

   (ii) Viviparity (viviparous animals)

   (iii) Ovoviparity (ovoviparous animals)

5. Animal sex hormones and gametes are produced in the ____________.

6. Although ___________ reproduction exists in animals, most animals reproduce ____________. (choices: sexual[ly] or asexual[ly])

7. Fill in the missing words from the list below: Animals in their ____________ advance through a number of stages: production of haploid __________, syngamy (fertilization) to form a __________ cell, _____________, _____________, ______________ (formation of organs), and ______________ and increase in size.

   Choices (not all used but used only once): differentiation, organogenesis, development, spermatogenesis, oogenesis, cleavage, diploid, haploid, gasification, gastrulation.

8. The overall process of forming gametes is called _________________. Sperm are produced specifically by the process of __________, and eggs by ______________.

9. The acrosome is the part of the sperm which bears _________ needed to penetrate the egg; the midpiece is crowded with _____________, organelles which power the movement of the flagellum. All that enters the egg, however is contained in the __________ of the sperm and consists of _____________. Actual movement of the flagellum is provided by __________in the _____________.

10. From one diploid parent cell, spermatogenesis produces ______ sperm and oogenesis produces _______ egg and ________ polar bodies.

11. In mammals, where are the mature sperm stored until ejaculation?

12. Which hormone is needed for spermatogenesis? Which cells in what organ produce this hormone?

13. Which pituitary hormones regulate the production of the hormone mentioned in the answer to question 12?

14. In humans, a woman's eggs are suspended in the prophase stage of _________ until puberty. Meiosis resumes shortly before_________, and the eggs progresses through the first stage. The _________ stage of meiosis is only completed after a sperm __________ the egg.

15. Hormones which affect the production of eggs and other aspects of the menstrual cycle in human females include (fill in the full name of the hormone following the given abbreviation or key word):

    (i) FSH ______________________________

    (ii) LH _______________________________

    (iii) corpus luteum ______________________

16. Male gonads are called ___________ and produce male sex cells or __________. Female gonads are called _________ and produce __________.

17. Fill in examples in the blanks. Fertilization can occur externally as in _________ or ___________, or internally as in _______________ or _______________.

18. Where does fertilization usually occur in the mammalian reproductive tract?

    a. vagina

    b. uterus

    c. oviduct (Fallopian tube)

    d. urethra

    e. ovary

19. What enters the egg during fertilization?

20. What immediately happens after the sperm nucleus enters the egg? What forms on the egg after it is fertilized, protecting it and preventing the intrusion of other sperm (called polyspermy)?

21. Three glands contribute fluid to make up the semen of mammals:

    (i) ______________________________

    (ii) ______________________________

    (iii) ______________________________

22. The fertilized egg is known as a _______________.

23. The division of the fertilized egg is know as ______________.

24. a. Place the following events in the right order, beginning with zygote:

    zygote, four-cell stage, blastula, differentiation, gastrula, two-cell stage, morula, formation of endo- and ectoderm, formation of mesoderm.

    b. when the blastula is formed from the morula, a cavity is created called the _________.

25. When is the archenteron formed? Where is it formed?

26. What will the blastopore eventually become?

27. In frogs, how are the animal and vegetal poles distinguished? Where does the grey crescent form and in what animals?

28. In the frog embryo, the animal pole cells will become __________. The vegetal pole cells will become ___________. Where does the mesoderm develop?

29. What is the amount and function of the yolk in an egg?

30. What are the three primary germ layers of the embryo.

31. Stem cells differentiate into _______ cells such as ___________ and _______________.

32. Shaping of the body form is called _____________, and proceeds by tissues _________________________, growing at different ___________, depositing material out side of the ______, and programmed cell ___________.

33. Ectoderm gives rise to ______________________________ (name four tissues).

34. Mesoderm gives rise to ______________________________ (name four tissues).

35. Endoderm gives rise to ______________________________ (name form tissues).

36. True or false; if false, explain why. Cells differentiate because they receive only the part of the DNA (genome) they need in order to function.

37. What is the tissue called which forms around what will become the central nervous system?

38. Briefly describe how a vertebrate eye forms.

39. Which organ system is the first to begin to develop?

40. What is the role of controlled cell death in development? Give an example.

41. What is meant by embryonic induction?

42. What is the name of the biologist associated with the concept of induction?

    a. Darwin

    b. Alexander

    c. Spemann

    d. von Baer

43. Testes are located outside of the body in a pouch called the __________ whereas the ovaries are located in the _____________.

44. Most nonhuman female mammals periodically enter a phase of fertility accompanied by sexual receptivity, called __________ or more technically, _____________.

45. Why are the testes located outside of the abdomen?

46. The process by which a prize bull can father thousands of calves is called ________________ ________________. (two words)

47. Are the following cells or structures located in the ovaries or testes?

    a. sertoli cells

    b. corpus luteum

    c. follicles

    d. leydig cells

    e. seminal vesicle

    f. vas deferens

    g. primary oocyte

    h. seminiferous tubules

48. True or false: A woman can get pregnant even if she does not have an orgasm during intercourse.

49. The fluid that causes the penis to become erect is ____________ trapped in spongy tissues.

50. The tip of the penis is known as the ________ penis.

51. The outer folds of tissue surrounding the opening to a woman's vagina are known as the ______________ __________ (two words); the inner folds are called the __________ _________ (two words). The small, sensitive organ which is analogous (and homologous) to the male's penis is the _____________.

52. Match the days of a 28-day menstrual cycle with these events:

    Events

    a. ovulation

    b. luteal phase in which the corpus luteum forms

    c. menstruation and sloughing of the uterine lining (endometrium)

    d. new uterine lining is made, follicle develops

    Days (approximate): day 1–day 5, day 6–day 13, day 14, day 15–day 28.

53. The onset of the menses (menstruation) in women is usually between the ages of ________ and ________.

54. A woman ceases ovulating and no longer menstruates when she reaches ___________ in the fifth or sixth decade of her life.

55. What extraembryonic membranes surround the embryo?

56. When does the human blastocyst become implanted in the wall of the uterus?

57. The growing embryo is fed through the ____________ attached to the wall of the uterus, through the blood vessels in the ___________.

58. What sort of materials can pass from mother-to-baby and vice versa?

59. Pregnancy may be divided into three ______________. During which one is the embryo/fetus most susceptible to chemicals which might cause birth defects?

60. Human pregnancy lasts for _______ weeks. ________ refers to the period before birth, __________ postnatal to the period of life following birth.

Check Yourself

1. Asexually, sexually, asexually, sexual **(Reproduction and development)**
2. genetic variation, diversity or variety **(Reproduction and development)**
3. False. For instance, there is a species of lizards consisting of parthenogenic females (the eggs develop without being fertilized). **(Reproduction and development)**
4. (i) eggs are laid outside the female's body—chickens, most reptiles, amphibians, and fish.

   (ii) embryo develops within the mother's body—almost all mammals.

   (iii) the eggs hatch within the oviducts—some snakes and fish.

   Note that the words themselves give clues—ovo refers to egg, vivi to live (as in live-bearing fish which hatch young within the oviducts), and what appears as a combination of the two others is called ovoviparity. Parity and parous refer to giving birth. **(Reproduction and development)**
5. gonads **(Reproduction and development)**
6. asexual, sexually **(Reproduction and development)**
7. development, gametes, zygote, cleavage, gastrulation, organogenesis, differentiation **(Reproduction and development)**
8. gametogenesis, spermatogenesis, oogenesis **(Gametogenesis)**
9. enzymes, mitochondria, nucleus, DNA, microtubules, tail **(Gametogenesis)**
10. four, one, three **(Gametogenesis)**
11. epididymis **(Gametogenesis)**
12. Testosterone is needed for sperm production, and is made in the Leydig cells in the testes. **(Gametogenesis)**
13. FSH (follicle stimulating hormone) and LH (lutenizing hormone) were initially discovered in female animals, hence their names. They are also produced in males. **(Gametogenesis)**
14. meiosis I, ovulation, second, fertilizes **(Gametogenesis)**
15. (i) FSH (follicle stimulating hormone)

    (ii) LH (lutenizing hormone)

    (iii) progesterone and some estrogen) **(Gametogenesis)**
16. testes, sperm, ovaries, eggs (ova) **(Gametogenesis)**
17. fish or amphibians, birds or mammals **(Fertilization)**

18. c. in the oviduct. The fertilized egg normally descends into the uterus; if it does not an ectopic or more specifically, a tubal pregnancy may result. **(Fertilization)**

19. The nucleus of the sperm enters the egg during fertilization. Since the sperm only contribute DNA to the union, the egg must supply organelles for the developing zygote. These are derived from the mother. As a result, DNA in mitochondria consists solely of maternal mitochondrial DNA. **(Fertilization)**

20. There is a depolarization of the egg plasma membrane immediately following the entry of the sperm. This fast block results from the opening of sodium channels as the two plasma membranes—egg and sperm—unite. Sodium ions rush out. Shortly thereafter, the slow block comes into play. Calcium ions are released, cortical granules in the egg rupture and release enzymes and other materials. The vitelline layer of the egg is thereby caused to swell with water and form the fertilization membrane, an effective barrier to additional sperm penetration. **(Fertilization)**

21. seminal vesicles, prostate gland, bulbourethral glands **(Fertilization)**

22. zygote **(Developmental mechanisms)**

23. cleavage **(Developmental mechanisms)**

24. (i) The process follows this order: zygote, two-cell stage, four-cell stage, morula (a ball of cells), blastula, gastrula, formation of the endo- and ectoderm, formation of mesoderm, differentiation.

    (ii) blastocoel **(Developmental mechanisms)**

25. The archenteron forms during the gastrula stage, as cells inward through the blastopore to form a cavity—imagine a partly filled volley ball which is punched in on one side to form a deep depression. The archenteron becomes the gut eventually, with the blastopore forming the anus in some animals, the deuterostomes. **(Developmental mechanisms)**

26. The blastopore later forms the mouth in protostomes or anus in deuterostomes. **(Developmental mechanisms)**

27. The dark part of the frog egg, where almost all cell contents except the yolk are found, is the animal pole. the opposite end is lighter in color and called the vegetal pole of the egg. In many amphibians, the point on the egg opposite where the sperm penetrated the egg forms the gray crescent. It marks what will later become the dorsal surface of the animal. **(Developmental mechanisms)**

28. Front of the animal, rear of the animal; between the ecto- and endoderm. **(Developmental mechanisms)**

29. Eggs vary greatly in the amount of yolk they contain. The sea urchin has very little yolk, fish more, and amphibians and birds much more yolk. Yolk is rich in many nutrients, especially fats, and provides energy to the growing embryo. **(Developmental mechanisms)**

30. The three primary germ layers are the endoderm, mesoderm and ectoderm. **(Developmental mechanisms)**

31. blood, RBC (red blood cells or erythrocytes), white cells (such as leukocytes and macrophages) **(Differentiation)**

32. morphogenesis, folding, rates, cell, death. Development moves forward by rearrangements of structures that were made in earlier stages. **(Developmental mechanisms)**

33. skin and hair, nerve tissues, lens of the eye, adrenal medulla. **(Differentiation)**

34. muscle and bone (connective tissue), kidneys and ureters, reproductive system, lymphatic system, circulatory system. **(Differentiation)**

35. gastrointestinal tract, linings of gastrointestinal and respiratory tract, liver and pancreas, thyroid, urinary bladder. **(Differentiation)**

36. False. Numerous experiments and assays of DNA have shown that all cells receive the complete genome (full complement of DNA) differentiation is accomplished by regulating which genes are turned on and off in a specific cell and tissue. **(Differentiation)**

37. The neural folds unite to form the neural tube, which in turn will surround the future spinal cord and brain. **(Organogenesis)**

38. At the front of the still-forming brain, two protuberances develop called the optic vesicles. These develop on stalks which push outward and become the optic nerves connecting the eyes to the brain. When the vesicle contacts surrounding ectodermal epithelium, it flattens and finally is pushed in, first forming a depression and then a cup-shaped cavity within the optic vesicle. The epidermal ectoderm gives rise to the transparent lens with the cupped vesicle behind it forming the eye. **(Organogenesis)**

39. The nervous system is the first of the major organ systems to start developing. **(Organogenesis)**

40. Cells and tissues that are formed during development and later are not required, are cast aside and resorbed through the process of programmed cell death—a built in control mechanism. Humans and ducks have webbed feet (sheets of skin between the toes) at some point in development. The web break down and disappear in humans, but not in ducks. Faulty genes in humans can result in a child born with webbed fingers and toes. **(Organogenesis)**

41. Briefly, embryonic induction is the influence of one tissue on another tissue during development—the differentiation of one tissue is directed by another. For instance, Hans Spemann was able to induce the formation of an eye on the belly of a salamander since the retinal tissue from the optic vesicle was able to induce the local ectodermal tissues to form a lens. **(Organogenesis)**

42. Spemann. The only other embryologist on the list is von Baer. **(Organogenesis)**

43. scrotum, abdomen **(Mammalian reproduction)**

44. heat, oestrus (estrus) **(Mammalian reproduction)**

45. The testes migrate down into the scrotum before birth. Sperm develop normally at a few degrees below body temperature and would not be able to do so if confined in the abdomen. High fevers can induce temporary sterility in men, supporting this argument. **(Mammalian reproduction)**

46. artificial insemination **(Mammalian reproduction)**

47. a, d, e, f, and h are located in the testes; b, c, and g are found in the ovaries **(Mammalian reproduction)**

48. True **(Human reproduction)**

49. blood. Blood trapped in the spongy tissues forms a hydraulic cylinder. **(Human reproduction)**

50. glans **(Human reproduction)**

51. labia major (labium majora), labia minor (labia minora), clitoris **(Human reproduction)**

52. a. day 14, b. day 15–day 28, c. day 1–day 5, d. day 6–day 13 **(Human reproduction)**

53. ten, sixteen **(Human reproduction)**

54. menopause **(Human reproduction)**

55. Amnion forms the first membrane, enclosing the embryo and becoming filled with amniotic fluid. The yolk sac is the next membrane, which once held yolk in our evolutionary past, but now functions as a site for red blood cells formation for the embryo. The chorion is the third membrane, which becomes part of the placenta, nourishing the embryo. the final membrane is the allantois. In birds and reptiles, this sac will hold embryonic waste products. In humans, like the yolk sac, it is responsible for making blood cells and later is incorporated into the urinary bladder. **(Human pregnancy)**

56. On about the sixth day following fertilization **(Human pregnancy)**

57. placental, umbilical cord **(Human pregnancy)**

58. Many substances, some necessary and required—food, salts oxygen, and water to the embryo, waste products from the embryo— but others may be harmful to the embryo—alcohol, ingredients of cigarette smoke, organic chemicals such as pesticides and solvents, and many medications. **(Human pregnancy)**

59. trimesters. During the earliest or first trimester the developing embryo is most at risk for damage due to materials crossing the placenta, or other effects such as ionizing radiation (e.g. X-rays). **(Human pregnancy)**

60. 38, prenatal, postnatal **(Human pregnancy)**

Grade Yourself

Circle the question numbers that you had incorrect. Then indicate the number of questions you missed. If you answered more than three questions incorrectly, you will then have to focus on that topic. (If a topic has less than three questions and you had at least one wrong, we suggest you study that topic also. Read your textbook, a review book, or ask your teacher for help.)

Subject: Animal and Human Reproduction and Development

Topic	Question Numbers	Number Incorrect
Reproduction and development	1, 2, 3, 4, 5, 6, 7	
Gametogenesis	8, 9, 10, 11, 12, 13, 14, 15, 16	
Fertilization	17, 18, 19, 20, 21	
Developmental mechanisms	22, 23, 24, 25, 26, 27, 28, 29, 30, 32	
Differentiation	31, 33, 34, 35, 36	
Organogenesis	37, 38, 39, 40, 41, 42	
Mammalian reproduction	43, 44, 45, 46, 47	
Human reproduction	48, 49, 50, 51, 52, 53, 54	
Human pregnancy	55, 56, 57, 58, 59, 60	

Animal Behavior and Communication

Brief Yourself

Behavior

Behavior is any reaction an animal makes to a stimulus. The behavior may be simple, complex, coordinated, instinctive, learned or a combination of these. Simple behaviors include those that are dependent on simple stimuli which elicit a specific, unvarying response. The stimulus is often referred to as a sign stimulus or more commonly, as a releaser. The programmed response is aptly called a fixed action pattern. An example is the greylag goose, in which a nesting female will roll a nearby egg—or even object that looks like an egg, such as a baseball—into her nest. If during this behavior the object to be rolled is removed, the goose will complete its action in the absence of the object.

Another example: If a simple behavior is genetically programmed and "hard-wired" into the nervous system, it is called an instinct or innate behavior. Not all instinctive behaviors are simple. Birds and spiders are genetically programmed to build nests and webs, which they do without learning from experience. Such more elaborate behaviors are thought to be based on a series of smaller, simple behaviors, which are done in sequence. Learning is not excluded from this process; it is just not required for the process to begin. Learning is the process by which an animal uses past experience to modify future behavior. Learning can be broken down into several topics, such as classical conditioning (in which an innate response becomes associated with a stimulus), or operant conditioning, where an animal learns a rewarding behavior by trial-and-error.

Many aspects of courtship are ritualized and fixed, species-specific behaviors that act as releasers for responsive behavior from the opposite sex. Social behavior is even more complex and can include hierarchy (dominance and submissiveness, reciprocal altruism and behavior based on kin-selection). Communication is an interesting aspect of social behavior, but may involve unexpected interaction with other species, such as predators alerted by signals. Eusocial insects (termites, ants, and some bees and wasps) have evolved social behavior and communication to a very sophisticated level.

It is common for animals such as birds, mammals, fish, and insects to spend one part of the year in northern breeding grounds, and retreat south with the coming of winter in order to secure food and protective habitat. These seasonal movements, or migrations, are demonstrated by monarch butterflies, ladybugs, songbirds, and caribou, among other animals.

Test Yourself

1. What is behavior?

2. What is a stimulus?

3. What is a taxis? List three examples of taxes.

4. A spider's web is an instance of ___________ behavior.

   a. learned

   b. instinctive (innate)

   c. taxis

   d. reflex

5. The scientist who discovered imprinting was:

   a. Lorenz

   b. Pavlov

   c. Tinbergen

   d. Hamilton

   e. Huxley

   f. von Frisch

   g. Wynne-Edwards

6. The orientation "dance of bees" is associated with the work of

   a. Lorenz

   b. Pavlov

   c. Tinbergen

   d. Hamilton

   e. Wallace

   f. von Frisch

   g. Wynne-Edwards

7. Self-sacrifice or altruism has been explained by ___________ as cooperative or reciprocal, that is, of benefit to each organism involved in the behavior.

   a. Lorenz

   b. Pavlov

   c. Tinbergen

   d. Hamilton

   e. Trivers

   f. von Frisch

   g. Wynne-Edwards

8. The existence of conditioned reflexes was established by the Russian physiologist ____________ .

   a. Lorenz

   b. Pavlov

   c. Tinbergen

   d. Hamilton

   e. Darwin

   f. von Frisch

   g. Wynne-Edwards

9. ____________ explained the sterile workers in insect societies (such as bees) as an example of kin selection, in which an organism behaves in a way so as to promote the inheritance of the copies of its genes which exist in another individual.

   a. Lorenz

   b. Pavlov

   c. Tinbergen

   d. Hamilton

   e. Darwin

   f. von Frisch

   g. Wynne-Edwards

10. _________________ is the main proponent of group selection, an alternative to explanations such as kin selection, and argues that animals can perform actions for the good of the group or species.

    a. Lorenz

    b. Pavlov

    c. Tinbergen

    d. Hamilton

    e. Wallace

    f. von Frisch

    g. Wynne-Edwards

11. When J.B.S. Haldane remarked he would lay down his life for two brothers or eight cousins, in his witty remark he was invoking a modern hypothesis concerning behavior that would be later known as:

    a. group selection

    b. imprinting

    c. conditioned response

    d. kin selection

    e. natural selection

12. A releaser is also know as a(n) ________________.

13. Fixed action patterns are behaviors which are

    a. unvarying

    b. repetitive

    c. inherited

    d. a, b, and c

    e. a and b

14. True or false: Imprinting is entirely an inherited response and does not involve learning.

15. The period in which imprinting can occur is called the ____________ period.

16. The red dot on a parent gull's beak is pecked at by a chick. The dot is an example of a(n) __________.

17. Behaviors cued to day/night cycles are called ____________ rhythms.

18. Briefly describe the migration of monarch butterflies.

19. In returning to the stream in which they were hatched, salmon are using ____________ cues.

    a. photoperiod

    b. chemical

    c. temperature

    d. lunar

    e. magnetic

20. The "pecking order" in chickens or submissive behaviors in wolves (and dogs) exist as part of a system of social ________________.

21. In singing, a male songbird is establishing its ________________.

22. Ants, termites, and some bees are __________ insects, in which the prefix ____- means "_______."

23. The normal development of singing in white-crowned sparrows, as studied by Marler, (pick ALL that apply)

    a. is entirely innate

    b. requires hearing the song of the adult males during a critical period (imprinting)

    c. requires the bird hear his own singing

    d. is partly based on inheritance

    e. requires the presence of a female mate

24. When Pavlov's dog salivated when a bell was rung, even though no food was present, this was an example of:

    a. an unconditioned response

    b. a conditioned response

    c. an unconditioned stimulus

    d. imprinting

    e. luck

    f. operant conditioning

25. When a person quickly withdraws his or her hand from a hot pot on the stove, this is an example of a _________.

26. The female stickleback will follow many types of _____ (a color) objects, apparently because these resemble the _____ (a color) abdomen of the male stickleback. At the nest, the male nudges her tail and causing her to release her _________. Later, the nest containing the fertilized eggs is guarded by the _________ (male or female). Stickleback courtship behavior was studied by ______________.

27. In a pack of wolves the dominant male is called the ________ male.

28. True or false: A male bird learns his song from his father or another adult male bird of the same species.

29. Which are more likely to migrate, insect- or seed-eating birds? Why?

30. Biologists who study animal behavior are called ____________.

31. A rat, put in the same maze again and again with food at the end as a reward, will navigate the maze in shorter and shorter periods of time. This is an example of ______________.

    a. an unconditioned response
    b. a conditioned response
    c. an unconditioned stimulus
    d. learning
    e. luck
    f. operant conditioning

32. Give three examples of animals who have a social dominance hierarchy.

33. Give an example of a simple, instinctive, inherited response seen in all humans.

34. A blue jay learns to avoid eating monarch butterflies after eating one, finding it distasteful or even becoming ill. This is an example of

    a. an unconditioned response
    b. a conditioned response
    c. an unconditioned stimulus
    d. imprinting
    e. luck
    f. operant conditioning

35. Behaviors which have __________ value tend to be perpetuated by natural selection.

36. What can account for the persistence of altruistic behaviors?

37. List four routes or means of animal communication, and give an example of each.

38. Jean-Henri Fabre, in the 19th century, discovered that if he put a female moth in a wire cage, males would collect around her, but not if he put her in a sealed glass container. Today, we know the males were attracted to the female by ____________.

39. A stimulus such as dawn that sets biological clocks referred to by a German word, __________, which means "time-giver."

40. In his book *Utopia,* Thomas Moore refers to ducks following a man about. He is most likely referring to an example of ______________.

41. A predator forms a ____________ of its prey, which helps it successfully hunt.

    a. search image
    b. protocol
    c. imprint
    d. lek

42. The bright red rump pads of a female baboon are an example of a __________ signal.

43. Is communication limited to that occurring between members of the same species?

44. Is territoriality only associated with birds?

45. The division of labor in termites and eusocial bee colonies is based on __________, in which reproductives and non-reproductives carry out physiologically assigned roles which limit breeding functions to a small segment of the population. The most likely explanation for this social organization is found in the sharing of ________ among colony members.

46. When males of such species as sage grouse gather during courtship about a common display area, the area is termed a ____________.

    a. lake
    b. lite
    c. lek
    d. lak

Check Yourself

1. Behavior is any response or reaction an animal makes to a stimulus. **(Elements and patterns of behavior)**

2. A stimulus is a change in the environment which does not provide energy for the animal's response. **(Elements and patterns of behavior)**

3. A taxis is a directional movement in response to a stimulus. Examples would be motile magnetotactic bacteria responding to a switch in the polarity of a magnetic field, the positive phototaxis of Euglena (a photosynthetic protist) in response to a light source, and chemotaxis, in which motile bacteria respond to the presence of organic molecules such as some sugars by moving up a concentration gradient to the source. **(Simple behaviors)**

4. instinctive or innate, that is, behavior based on genes the spider inherited **(Complex behaviors)**

5. a. Konrad Lorenz **(Complex behaviors)**

6. f. Karl von Frisch **(Complex behaviors)**

7. e. Robert Trivers **(Social behavior)**

8. b. Ivan Pavlov **(Simple behaviors)**

9. d. W.D. Hamilton. This explanation is an alternative to the group selection hypothesis. **(Social behavior)**

10. g. V.C. Wynne-Edwards supported this hypothesis, which is not accepted by most biologists. **(Social behavior)**

11. d. Haldane was indicating that even if he lost his life and did not reproduce, this number of brothers or cousins would transmit enough of the genes he and they shared in common to increase his fitness (increase the number of his genes in their descendants in subsequent generations). **(Social behavior)**

12. sign stimulus. **(Simple behaviors)**

13. d. a, b, and c **(Elements and patterns of behavior)**

14. False. While imprinting is genetically based, it requires a learning response to a stimulus or stimuli presented during the critical period for that imprinted behavior. **(Learning)**

15. critical. This is the period in an animal s life during which an imprint must be established. **(Complex behaviors)**

16. releaser or sign stimulus **(Complex behaviors)**

17. circadian These are rhythms based on an approximately 24-hour clock. **(Complex behaviors)**

18. At the end of each summer, newly-emerged adults migrate from northern areas across North America to regions in Mexico, Florida and California to spend the winter. In the spring, the butterflies migrate north, and after several generations there, begin the southward migration in the fall. **(Migration)**

19. b. chemical cues which are specific to the stream in which they were born. **(Migration)**

20. dominance. **(Social behavior)**

21. territory. More specifically, it is establishing its breeding territory, to which it attempts to attract a mate and warn away competing males. **(Social behavior)**

22. eusocial, eu, true. **(Social behavior)**

23. b, c, and d are true and apply to the normal development of song. The normal song will not be developed if the bird is deaf and cannot hear the song of another male or its own singing, or if it is not exposed to the singing of a male adult during the critical imprinting period.The behavior is thus not entirely innate (instinctive), and at no time is the presence of a female required for song development. **(Complex behaviors)**

24. b. This is because the dog learned to associate the sound of the bell with food. **(Simple behaviors)**

25. reflex, an action carried out in the nerves and spinal cord **(Simple behaviors)**

26. red, red, eggs, male, Tinbergen. The female will respond to red objects of various sizes and shapes.This courtship behavior is an example of a series of simple behaviors linked together to form a more complex behavior. **(Courtship and mating)**

27. alpha **(Social behavior)**

28. True, in the sense that the bird must be exposed to a male singing during the critical period of song imprinting for normal song development. **(Complex behaviors)**

29. The supply of insects in northern regions is more seasonal than the supply of seeds, making it more likely that an insect eater will need to migrate. **(Migration)**

30. ethologists. This word can be easily confused with the name for scientists who study human culture, ethnologists. Recall this latter word has the same root as the word "ethnic." **(Elements and patterns of behavior)**

31. f. operant conditioning. That is, a trial and error form of learning, in which an animal learns to make a non-innate response to a stimulus through the use of rewards. For example, a pigeon can be taught to peck at a square figure rather than a circle or triangle in order to get a reward of food. B.F. Skinner, a psychologist, is widely associated with this term. **(Learning)**

32. Chickens show dominance in their "pecking order," dogs and wild canids like wolves have dominant and submissive pack members, and a dominance hierarchy is evident in a number of primates, such as baboons. **(Social behavior)**

33. Smiling is one good example, or as in all mammalian young, instinctive sucking when presented with a nipple or nipple-like object. **(Simple behaviors)**

34. b. The blue jay learns to associate the unpleasant experience with the appearance of the monarch butterfly. **(Learning)**

35. adaptive. These are behaviors which tend to promote the production and survival an increased number of offspring. **(Elements and patterns of behavior)**

36. Altruistic behaviors will be perpetuated if they are adaptive and thus selected for in evolution (see answer to question 35). **(Elements and patterns of behaviors)**

37. Examples would include:

    a. tactile, as when the male stickleback nudges the female's abdomen

    b. chemical, such as pheromones which signal a female in heat, or in the urine of an animal as a territorial scent marker

    c. visual, when a worker bee indicated the direction and distance of a food source to other workers by dance movements and tail-wagging; or the display of colored plumage in many birds; or the light signals of fireflies

    d. auditory, as when a bird sings to establish a territory and a mate **(Communication)**

38. pheromones **(Communication)**

39. zeitgeber **(Complex behaviors)**

40. imprinting **(Learning)**

41. a. search image **(Learning)**

42. visual **(Communication)**

43. No. A frog-eating bat can home in on an amphibian's calls as a hunting strategy—the communication of the frog is exploited by a predator. **(Communication)**

44. No. Wolf packs, lion prides, and males of the deer family can all determine and protect territories. Territoriality is even seen in insects such as dragon flies, where males protect their sections along a stream bank from competing males. **(Social behavior)**

45. caste, genes. **(Social behavior)**

46. c. lek **(Courtship and mating)**

Grade Yourself

Circle the question numbers that you had incorrect. Then indicate the number of questions you missed. If you answered more than three questions incorrectly, you will then have to focus on that topic. (If a topic has less than three questions and you had at least one wrong, we suggest you study that topic also. Read your textbook, a review book, or ask your teacher for help.)

Subject: Animal Behavior and Communication

Topic	Question Numbers	Number Incorrect
Elements and patterns of behavior	1, 2, 13, 30, 35, 36	
Simple behaviors	3, 8, 12, 24, 25, 33	
Complex behaviors	4, 5, 6, 15, 16, 17, 23, 28, 39	
Social behavior	7, 9, 10, 11, 20, 21, 22, 27, 32, 44, 45	
Learning	14, 31, 34, 40, 41	
Migration	18, 19, 29	
Courtship and mating	26, 46	
Communication	37, 38, 42, 43	

Ecology, Cycles, and Populations

Brief Yourself

A brief timetable reviewing the origin and evolution of life on earth:

Age of the earth: 4.5 billion years ago (bya)

Life first appeared: 3.5 bya

Precambrian era: 590–4,500 million years ago (mya). Prokaryotes and eukaryotes, multicellular life (700 mya). Major geological changes, including formation of mountains and an atmosphere with free oxygen

Paleozoic era: 590-248 mya. Invertebrates, fungi, land plants, fishes, first land vertebrates (about 350 mya), amphibians and reptiles, gymnosperms.

Mesozoic era: 248-65 mya. Dinosaurs rise and fall. Period starts with forests dominated by gymnosperms, ends with rise of flowering plants. Mammals and birds appear.

Cenozoic era: 65 mya to present. Mammals and flowering plants increase in diversity and abundance. Continents drift to modern-day arrangement. Earliest primates arise at beginning of Cenozoic; human ancestors (hominids) found 5 mya or earlier.

An ecosystem embraces the entire living and nonliving environment in an area. Ecosystems differ in the number of species (called diversity or species richness) and the number of individuals of each species (sometimes called species equatability) present. For instance, tropical ecosystems are characteristically rich in both ways; while far north ecosystems have a sparse number of species and fewer individuals. Species and individuals compete for resources, many of which are relatively scarce. Stability varies. The biologic makeup changes over time (succession, disruption, etc.) in ecosystems as organisms jockey for positions in which to make a living (niches) and live (habitat) and the climate varies. Abiotic factors are part of the ecosystem. Ecosystems will develop differently depending on the amount of sunlight and precipitation received, the make-up of the soil and the temperature. In watery ecosystems, oxygen can play a critical determining role. Biomes are another way of looking at ecosystems. Biomes are regions characterized by certain distinct types of dominant vegetation. Eight types of biomes are widely recognized: Tropical Rainforest, Savanna, Desert, Chaparral, Temperate Grasslands, Temperate Deciduous Forests, Taiga, and Tundra.

Bio- and Geochemical Cycles

Chemical elements and compounds (for example: nitrogen, carbon, phosphorus, oxygen, and water) are cycled through the biosphere, from environment to organism and back again, providing these essential resources to generation after generation of organisms. A little thought given to the chemical composition of compounds of biological importance, such as proteins and nucleic acids, will demonstrate the vital role played by these elements.

Energy and Trophic Levels

In an ecosystem, organisms are linked to each other and their environment by the flow of energy and material resources. The only sources of energy for organisms on earth is sunlight and geochemical energy from vents in the ocean floor. Almost all energy originates with photosynthetic autotrophs (primary producers) which capture the sun's energy and plug it into the ecosystem. These are fed on by herbivorous primary consumers (as diverse as cattle and insects), which in turn feed secondary consumers (primary carnivores such as some fish and insects) and finally tertiary consumers (secondary carnivores, such as birds which eat fish fed by other fish). Food webs or chains are another way of describing and envisioning energy flow through an ecosystem—a matter of who eats whom. Animals commonly obtain energy from several trophic levels and this makes for complex interrelationships. Energy relationships can be quantified. Gross primary productivity refers to the total amount of biomass produced by photosynthesis in a given time, while net primary productivity is the gain realized after the metabolic costs of the plants are subtracted.

Ecological succession

Ecological succession is a series of changes in the ecology of an area over time. In general, these changes reflect the growing, dynamic complexity of relationships of the organisms with each other and their environment.

Test Yourself

1. The earth is about how many billion years old?

2. Life first appeared on earth ________ billion years ago.

3. During the Precambrian era, major events in the history of life included which of the following?

   a. Prokaryotes and eukaryotes appeared and an atmosphere with free oxygen developed.

   b. Forests dominated by gymnosperms, mammals and birds appear.

   c. Invertebrate animals, fishes and first land vertebrates appear, along with gymnosperms.

   d. Mammals and flowering plants increase in abundance, and primates appear. Flowering plants become a dominant form of plant life.

4. During the Paleozoic era, major events in the history of life included which of the following?

   a. Prokaryotes and eukaryotes appeared and an atmosphere with free oxygen.

   b. Forests dominated by gymnosperms, mammals and birds appear.

   c. Invertebrate animals, fishes and first land vertebrates appear, along with gymnosperms.

   d. Mammals and flowering plants increase in abundance, and primates appear. Flowering plants become dominant form of plant life.

5. During the Mesozoic era, major events in the history of life included which of the following?

   a. Prokaryotes and eukaryotes appeared and an atmosphere with free oxygen.

   b. Forests dominated by gymnosperms, mammals and birds appear.

c. Invertebrate animals, fishes and first land vertebrates appear, along with gymnosperms.

d. Mammals and flowering plants increase in abundance, and primates appear. Flowering plants become dominant form of plant life.

6. During the Cenozoic era, major events in the history of life included which of the following?

   a. Prokaryotes and eukaryotes appeared and an atmosphere with free oxygen.

   b. Forests dominated by gymnosperms, mammals and birds appear.

   c. Invertebrate animals, fishes and first land vertebrates appear, along with gymnosperms.

   d. Mammals and flowering plants increase in abundance, and primates appear. Flowering plants become dominant form of plant life.

7. Life originated on/in the __________. (land, oceans)

8. The oceans are divided into three zones (habitats): ____________ (deep, cold and dark regions inhabited by highly specialized life forms), ____________ (approximately the top 50–100 meters that sunlight to some degree penetrates—out of an average ocean depth of 4 km—inhabited by autotrophic plankton) and _________ shallower, coastal waters that lie on top of the continental shelves, including fish, coral reefs and tidal organisms).

   Fill in each of the blanks with one of these terms (not all are used): limnetic, abyssal, neritic, littoral, profundal, surface.

9. Algae, small animals, and other free-floating organisms found in the oceans are termed ________, while _________ organisms inhabit the bottom areas. The water making up the oceans as a whole is called the ________ province, while the sediments and other bottom materials form the _________ province.

10. The western sides of Northern hemisphere land masses are warmer than the east coast shores because of the influence of _________ _________. (two words)

11. _____________ are places where the freshwater of a _________ or __________ flows into the saltwater of the ocean.

12. Lakes and ponds have three zones: (i) ___________ (from the shore to about 10 meters in depth); (ii) __________ (open water of a lake); and (iii) ____________ (deep areas beyond significant without significant penetration of light). Additionally, the penetration of light can be used to divide lakes into zones. _________ zones exist where light can penetrate to a degree adequate to allow photosynthesis, while in _________ zones photosynthesis is not possible. In the aphotic (deeper) zone, living organisms are confined to decomposers and those that feed on them.

Fill in each of the blanks with one of these terms (not all are used): limnetic, abyssal, neritic, photic, littoral, profundal, aphotic, littoral, surface.

13. What limits the growth of organisms and populations in water ecosystems? What element can play a critical role in water habitats?

14. Nutrient-poor lakes are ____________, while __________ lakes or ponds are nutrient rich, sometimes too much so due to fertilizer runoff from cultivated land.

15. Which of the following are parts of an ecosystem? (List all that apply.)

   a. geology of an area (rocks and earth)

   b. vegetation

   c. bodies of water

   d. animal inhabitants

   e. factors such as amount of sunlight, oxygen concentration (e.g., in water), temperature range and average

Questions 16–23. Match the correct biome with the description

Possible choices (each used once): tropical rain forest, savanna, desert, chaparral, temperate grassland, temperate deciduous forest, taiga, tundra

16. Can be found above the treeline; deeper ground never thaws completely.

17. Poor soil as most nutrients are locked up in biomass; rich in individuals and number of species.

18. Variable, low annual rainfall; low population densities of organisms with unique adaptations.

19. The Great Plains are an example.

20. Frequent fires, moderate, wet winters, and hot dry summers.

21. Characterized by conifer forests; short summers but long days suggest the latitude.

22. Herds of grazing herbivores, and clumps of trees, are two predominant features.

23. Found only in the Northern Hemisphere, with a variety of vegetation layers and animal life.

24. Most nitrogen is found in the ___________. About ___ percent of nitrogen naturally made available to organisms is through the action of ________ - _________ (two words, hyphenated) bacteria or closely related organisms; the other 10 percent through ___________.

25. Carbon is ________ in the form of carbohydrates by __________ __________ (two words) and is returned to the atmosphere through the __________ of organisms.

26. Burning fossil fuels releases what? What is one potential problem associated with this release

27. What one element of major importance to life is found in the earth s crust and initially released to organisms by weathering?

28. One element appeared in the course of the evolution of life on earth, and profoundly changed its direction. Which one?

29. a. What is a watershed?

    b. What effect does rainfall have on vegetation ?

Questions 30–35. Match the description or example with the correct term.

Possible answers (all used, and only once): detritivores, primary carnivores, secondary carnivores, secondary consumers, decomposers, primary producers.

30. Bacteria and fungi

31. Photosynthetic autotrophs

32. Eat primary consumers

33. Primary carnivores are this type of consumer

34. Earthworms

35. Birds which eat fish fed by other fish

36. Distinguish gross from net primary productivity.

37. How does primary succession differ from secondary succession?

38. What form of succession follows the abandonment of farmland?

39. Give an example of:

    a. a pioneer organism

    b. type of trees predominating in a climax forest.

40. a. What is a population?

    b. What is a community?

41. What are the three ways in which organisms can be distributed in a population's region?

42. What are some density-dependent factors affecting population size?

43. What are some density-independent factors affecting population size?

44. Population size which increases slowly at first, and then takes off rapidly until the carrying capacity is reached is termed _________.

45. If natural selection is constantly producing better predators, why don't prey species become extinct?

46. Predators on plants are called ________________.

47. List some factors which determine the carrying capacity of an environment.

48. Indicate whether a type I, II, or III curve would be expected with each of the following organisms:
    a. small prey animals
    b. bacteria
    c. elephants, whales, and humans
    d. most invertebrates
    e. many song birds

49. Indicate if the following characteristics (or organisms) are generally associated with r-selected or K-selected organisms.
    a. large clutches of young
    b. small clutches of young
    c. later reproductive maturity
    d. stable environments
    e. unstable, changing environments
    f. many small offspring
    g. extended parental care
    h. elephants
    i. shrimp

50. Give two examples of plant defenses against herbivory.

51. Give an example of a herbivore's coevolutionary adaptation to counter plant defenses.

52. Coloration which has a protective role is called ____________ coloration
    a. apposite
    b. apoplectic
    c. Batesian
    d. aposematic
    e. Mullerian

53. When two or more inedible species resemble each other, this is an example of __________ mimicry.

54. The resemblance of the viceroy to the monarch butterfly is an instance of ___________ mimicry.

55. Monarch butterflies derive their toxin from eating __________ as larvae.

56. Classify each of the following as (i) commensalism, (ii) mutualism, or (iii) parasitism
    a. bacteria in an infected finger
    b. the bacteria normally found in an animal's gut
    c. bacteria and protists in the gut of a ruminant
    d. fleas on a dog
    e. barnacles on whales
    f. birds which pick mites and other parasites off of other animals (like hippos)
    g. malaria in humans or birds

57. Give an example of coevolution.

58. Give an example of pollution.

59. Discuss the various ways in which organisms may interact, including competition for scare resources and the competitive exclusion principle.

Check Yourself

1. 4.5. **(Life on earth)**

2. 3.5. **(Life on earth)**

3. a. **(Life on earth)**

4. c. **(Life on earth)**

5. b. **(Life on earth)**

6. d. **(Life on earth)**

7. oceans. **(Life on earth)**

8. abyssal, surface, neritic. Limnetic, littoral, and profundal refer to freshwater zones. Today, oceans cover most of the earth and contain life that is both diverse and abundant. The oceans are divided into three zones (habitats): abyssal (deep, cold, and dark regions inhabited by highly specialized life forms), surface (approximately the top 50–100 meters that sunlight to some degree penetrates—out of an average ocean depth of 4 km—inhabited by autotrophic plankton) and neritic (shallower, coastal waters that lie on top of the continental shelves, including fish, coral reefs and tidal organisms). The water making up the oceans as a whole is called the pelagic province, while the sediments and other bottom materials form the benthic province. **(Water: oceans and freshwater)**

9. Plankton, benthic, pelagic, benthic. Free-floating life is termed plankton, and consists of algae, a diversity of protists, shrimp and other organisms. Benthic organisms are generally sessile and inhabit bottom areas. These include plentiful bacteria and fungi, sponges, molluscs, crustaceans, sea stars (starfish), and fishes. The tidal zones and seashores contain many organisms adapted to the currents and alternating periods of wet-and-dry in the intertidal zone, including clams and mussels, crustaceans, brown and other algae. The parts that are always submerged in water—the subtidal zone—contains similar life, plus kelp in such abundance as to be likened to forests. Seashores may be muddy, rocky or sandy, and will have flora (plants) and fauna (animals) adapted to these conditions. **(Water: oceans and freshwater)**

10. ocean currents. Ocean currents have a major influence on climate, producing coastal waters and land regions which are cooler or warmer than would be predicted on the basis of latitude alone—for example, the western sides of Northern hemisphere land masses are warmer for this reason than the east coast shores. **(Water: oceans and freshwater)**

11. estuaries, river, stream. Estuaries are places where the freshwater of a river or steams flows into the ocean. The water is brackish to a varying extent; where a great river like the Amazon meets the sea, salinity is affected for about 100 miles from the mouth of the river. Nutrients suspended or dissolved in the freshwater, which was washed from the land, encourages the abundance of life found in such areas. **(Water: oceans and freshwater)**

12. littoral, limnetic, profundal, photic, aphotic. Freshwater can be found in lakes, ponds, rivers, and streams, and in temporary or permanent bodies. Lakes and ponds have three zones: (i) littoral (from the shore to about 10 meters in depth); (ii) limnetic (open water of a lake); and (iii) profundal (deep areas beyond significant without significant penetration of light). Additionally, the penetration of light can be used to divide lakes into zones. Photic zones exist where light can penetrate to a degree adequate to allow

photosynthesis, while in aphotic zones photosynthesis is not possible. In the aphotic zone, living organisms are confined to decomposers and those that feed on them. **(Water: oceans and freshwater)**

13. The growth of individuals and populations in aquatic or marine ecosystems is limited by the availability of nutrients (as opposed to living space, for instance). Oxygen levels can also be of critical importance. **(Ecosystems)**

14. oligotrophic, eutrophic. Growth of organisms in water is limited by available nutrients. The oceans, for all their volume, are relatively nutrient-poor, and thus have a lower productivity per square mile than most land. In freshwater, a similar situation exists. However, nitrogen and phosphorus levels can sometimes be augmented by nutrients washed from cultivated land. This can lead to overgrowth of plants, or algal bloom which chokes out many other life forms.

    Eutrophic lakes are rich in organic and mineral nutrients, while oligotrophic lakes are nutrient-poor. **(Water: oceans and freshwater)**

15. All of these answers describe parts of ecosystems, which are the sum total of biological and abiotic aspects of a region. **(Ecosystems)**

16. Tundra. Theses can be arctic or alpine (above the treeline). Amount of rainfall/snowfall is low. The ground below one meter or less remains permanently frozen (permafrost), limiting how deeply roots of plants can grow. Scattered small trees. Inhabited by grazing animals like caribou and small mamma's adapted to the cold climate. **(Biomes)**

17. Tropical rain forest. These are located near the equator; warm, wet regions, highly productive and biologically diverse, canopied forest, but with poor soil as most nutrients are locked up in vegetation **(Biomes)**

18. Deserts. These are biomes characterized by variable, low annual rainfall. Plants and animals are found at low population densities, and exhibit adaptations for extreme dryness. Rainfall dictates when plants grow and reproduce. Daily temperature may vary widely. **(Biomes)**

19. Temperate grassland. These are open regions such as the Great Plains, which basically are temperate versions of the tropical savannas. Winters tend to be cold and long, and fire and drought limit the number of trees. Grazing animals are common and can occur in great numbers as once did the American bison. **(Biomes)**

20. Chaparral. These are coastal areas with moderate, wet, winters and dry, hot summers. Fires occur often, and plants show adaptations to this, including seeds that require exposure to fire to germinate. Found in California and the Mediterranean. **(Biomes)**

21. Taiga. These are found at higher latitudes (and higher temperate altitudes), colder and drier than temperate forests, favoring conifer forests. Long days make up for short summers. Inhabited by large, browsing herbivores and carnivores. **(Biomes)**

22. Savanna. These are open grasslands with a scattering of trees that form a transition from tropical rain forest to desert; have distinct seasons differing in temperature and rainfall. Often inhabited by vast herds of grazing herbivores. **(Biomes)**

23. Temperate deciduous forest. Thesea are limited to the Northern hemisphere. Greater rainfall, distributed throughout the year, allows the establishment of forests with multiple layers of vegetation and a wide variety of animal life. **(Biomes)**

24. atmosphere (about 78 percent to be exact), 90 percent, nitrogen-fixing, lightning. Nitrogen makes up 78 percent of the earth's atmosphere, but these riches are unavailable to most organisms in this form. Ultimately, most organisms must obtain their nitrogen through the activities of the several processes by which atmospheric nitrogen is fixed. About 90 percent of the nitrogen made available is through the agency of specialized organisms. Nitrogen-fixing bacteria convert atmospheric nitrogen to ammonia. In turn, other bacteria convert ammonia to nitrites (by *Nitroso* bacteria) and nitrates. Some transformation into ammonia is carried out first by blue-green algae (cyanobacteria) and related organisms (bacteria) found in the earth and water. Some of these nitrogen-fixing organisms live in symbiosis with legumes, some trees such as alders, and grasses. Decomposers break down nitrogenous wastes and dead organisms into ammonia, returning nitrogen to the air or into the nitrogen cycle to become nitrites and nitrates. (See symbiosis section, below). Denitrifiers also convert nitrates or nitrites to nitrogen. About 10 percent of nitrogen is fixed by lightening—nitric acid is formed and falls to the ground in rain.While plants can make direct use of the products of microbial nitrogen fixation, animals cannot. Animals must eat plants or plant-eaters to get nitrogen. **(Bio- and geochemical cycles)**

25. fixed, photosynthetic autotrophs, respiration. Carbon dioxide is released into the air by the respiration of organisms, volcanic action, and in modern times, in significant amounts by the burning of wood and fossil fuels by humans. Autotrophs fix carbon in the form of carbohydrates during photosynthesis, and make them available to herterotrophic organisms who eat plants or plant-eaters. Carbon reservoirs include (in order of importance); oceans, soil, atmosphere, organisms. **(Bio- and geochemical cycles)**

26. Carbon dioxide. This is the principal gas of the greenhouse effect which may be warming the earth. **(Bio- and geochemical cycles)**

27. The sentence best describes phosphorus as it is of "major importance"; other more minor elements are also released to organisms in this manner. Phosphorus is found in mineral form in the earth's crust, and is slowly made available through weathering. As in the case of nitrogen, plants are primarily responsible for making phosphorus available to other organisms as they can assimilate inorganic phosphorus. Wastes and dead organisms return them to the cycle to be used again. Many other mineral nutrients follow cycle like that of phosphorus. **(Bio- and geochemical cycles)**

28. Free oxygen was lacking in the original atmosphere under which life originated and initially evolved on earth. It was later produced in large quantities by photosynthetic autotrophs and made possible the now predominate, oxygen-dependent forms of life. Oxygen (free oxygen) levels in the atmosphere have varied tremendously during the history of life. Photosynthetic organisms produce oxygen as a by-product of photosynthesis, and which is used in cellular respiration. The historical rise of such organisms profoundly changed the composition of the air and dramatically altered the course of evolution. **(Bio- and geochemical cycles)**

29. a. A watershed is an area which collects rainfall for and channels it to a stream, river, pond or lake.

    b. Rainfall often determines when vegetation reproduces and grows during the year. The water cycle is critical to life. Most water is present in the free state, and in circulation between organisms, oceans, atmosphere and even the earth. Water can be found on the surface or in underground repositories called groundwater. Areas that channel precipitation through the soil or as runoff into one stream, river or lake are called watersheds. **(Bio- and geochemical cycles)**

30. decomposers. Decomposers (bacteria and fungi) augment the activity of consumers by breaking down wastes and dead organisms, in effect recycling them. Detritivores eat detritus—the small organic bits in the soil that are eaten by earthworms, for example. **(Energy and trophic levels)**

31. primary producers **(Energy and trophic levels)**

32. primary carnivores **(Energy and trophic levels)**

33. secondary consumers **(Energy and trophic levels)**

34. detritivores **(Energy and trophic levels)**

35. secondary carnivores. Almost all biological energy originates with photosynthetic autotrophs (primary producers) which funnel the sun's energy into the ecosystem. These are fed on by primary consumers (as diverse as cattle and insects), which in turn feed secondary consumers (primary carnivores such as some fish and insects) and finally tertiary consumers (secondary carnivores, such as birds which eat fish fed by other fish). Useful energy is lost as heat at each step of the trophic pyramid (so-called as there are many primary producers at the bottom, ranging to fewer and fewer consumers as one goes up through the trophic levels). The pyramid rests on a foundation of autotrophic activity; all other organisms, other than chemoautotrophic bacteria, are heterotrophs and directly or indirectly dependent on that first layer. Food webs or chains are another way of describing and envisioning energy flow through an ecosystem—a matter of who eats whom. Animals commonly obtain energy from several trophic levels and this makes for complex interrelationships. Note: confusion over these terms, and how the fact that two terms may be applied to a trophic level (primary carnivore but secondary consumer) may be remedied by drawing diagrams of the pyramid of trophic levels. **(Energy and trophic levels)**

36. Gross primary productivity is the total amount of biomass made by photosynthesis in a given time, while the net primary productivity is the same number minus the metabolic costs that plants incur during that time. **(Energy and trophic levels)**

37. Primary succession starts with a new, barren landscape, uncolonized by life. Secondary succession is the return of an area to a natural state after a disturbance. **(Succession)**

38. Secondary. Primary succession occurs when pioneer species enter a new, barren areas, for example, lichen forming colonies on rocks. Conditions (such as the slow formation of soil) improve for other organisms as a by-product of the activities of pioneer species. In secondary succession, an area returns to its original, natural state when an intrusion ceases; for example, when farmland is abandoned it recovers. In this specific instance, the following years will see the appearance of shrubs, then pines and finally hardwood tree species. A final form in succession is termed the climax community, where a state of equilibrium has been reached and will be maintained in the absence of disturbing forces. The make up of a climax community, like all other stages of succession, varies with geography and climate. **(Succession)**

39. a. lichen, b. hardwoods **(Succession)**

40. a. A population is a group of interbreeding organisms. Population size, density, and distribution (how individuals are arranged or spaced about the habitat—clumped, random or uniform patterns) as well as age structure (how many individuals in each age class) characterize a population. For instance, trees on the savanna are found in distinct groupings (clumped), and not even distributed. Foraging for resources such as food or shelter can cause individuals to move apart form other members of the population (e.g., a predator like a lion, born into a family group may migrate to find more prey).

    b. Biologists refer to communities when they want to discuss the relationship between the various population in a habitat. Each species is envisioned as having its own niche or unique physical and biological living space in the habitat. **(Populations and communities)**

41. clumped (trees on the savanna), random (individual softwoods in a hardwood forest), and uniform (distribution of some weeds which densely populate a lawn) **(Populations and communities)**

42. Increasingly limited resources, predators and parasites. **(Regulation of population size)**

43. Unusual or bad weather, fires, floods and other incidents that are not more or less likely to happen due to changes in the population density of a species. Population size is affected by two types of factors: (i) Density-independent factors or controls influence population size regardless of how many members of population are found in a given area. Bad weather, late springs or early falls, and chance events like floods or fires fall into this class. (ii) Density-dependent factors are those that slow a populations growth as the number of individuals in an area increase; competition for limited resources is heightened, and predators and parasites thrive. **(Regulation of population size)**

44. logistical. Populations tend to change in size over time in a predictable manner. In a logistic pattern, population size increases slowly and then reaches a period of rapid expansion, followed by a leveling off of number as the carrying capacity (the number which can be indefinitely sustained by resources in this area) is reached and exceeded. Limiting factors which determine the carrying capacity include space, food and various nutrient resources. Other patterns can be (i) near-logistic as when nonnative animals are first introduced onto an island and initial population growth is explosive; (ii) boom-and-bust as a result of overexploitation of resources and a subsequent decline in numbers as from starvation; (iii) seasonal, typically in which populations of plants or invertebrates rise and fall with seasonal changes. **(Regulation of population size)**

45. As predators are getting more efficient (e.g., faster) they cause selection in favor of the members of prey species most able to resist their new advantages. That is, as one gets better, so does the other. Predators and their prey are involved together in a complex evolutionary dance. As predators evolve better hunting traits, the prey is simultaneously selected for heightened antipredation traits. Thus, while prey populations may fluctuate, predators and their prey maintain reasonably stable populations. **(Interactions: competition, predation and symbioses)**

46. Herbivores **(Interactions: competition, predation and symbioses)**

47. Climatic factors (amount of sunlight, temperature, rainfall amount and pattern), other animals or plants (abundance of prey), weather, specific deficiencies in the soil, etc. **(Regulation of population size)**

48. a. II, b. III, c. I, d. III, e. II

    The type I curve applies to animals such as elephants and humans, in which individuals live a long time on average, and invest a great deal in parental care of offspring. The type II curve is characteristic of small prey animals, such as many common song birds. Mortality is fairly evenly distributed over time, meaning any one individual runs a fair chance of dying at any age. The type III curve describes a population in which many offspring are produced, and accompanied by high mortality of the young. Included in this group are many invertebrates, fish, insects, bacteria, fungi, and plants. Little or no effort or resources are invested in parental care. **(Life history)**

49. a. r, b. K, c. K, d. K, e. r, f. r, g. K, h. K, i. r.

    r- and K-selection are really generalizations about life history traits, and specifically refer to reproductive strategies. ("r" is adopted for this use since in mathematical equations it represents the rate of population growth, and "K" the carrying capacity of the environment.) r-selection is a consequence of living in ever-changing environments; for organisms which successfully live under such conditions, the uncertainty favors rapid population increase. r-selection fosters early reproductive maturity, the production of many, small offspring, and frequent broods, but little or no parental care once eggs are laid, for instance. r-selected populations tend to show great fluctuation in numbers (e.g., aphids). K-selection occurs in more stable and certain environments. Successful organisms in this case tend to have fewer and larger offspring, a stable population at or near the carrying capacity of that environment, and may invest considerable resources in ensuring the survival of the offspring (e.g., elephants). **(r- and K-selection)**

50. spines or thorns, secondary compounds like alkaloids or tannins. Plants as wells as animals have predators, that is, herbivores. Just as in the case of animal prey, plants have also evolved a range of anti-predator devices: (i) specialized organs such as thorns, prickles, silica particles (horsetails); (ii) woody tissue resistant to attack and digestion; (iii) chemical defenses, consisting of secondary (metabolic) compounds produced for protection. Some compounds are noxious and may produce burning or other unpleasant effects (oxalic acid in rhubarb leaves, capsicum in hot peppers, tannins in nuts), others may be poisonous and cause vomiting or even death (nicotine, or cardiac glycosides in milkweed). **(Interactions: competition, predation and symbioses)**

51. thick mouth tissues. As with carnivores, herbivores in turn have evolved ways around plant defenses—thick insensitive mouth tissues against spines, or the ability to digest woody tissue in termites (using protozoans in the gut). **(Interactions: competition, predation and symbioses)**

52. d. aposematic **(Interactions: competition, predation and symbioses)**

53. Mullerian **(Interactions: competition, predation and symbioses)**

54. Batesian **(Interactions: competition, predation and symbioses)**

55. milkweed. Monarch butterflies, feeding on milkweed as larvae, incorporate the cardiac glycosides of the plant into their own body, which makes them distasteful butterflies to predators. Such animals often signal potential predators by advertising their defense through distinctive aposematic (protective) coloration Building yet another layer on this series of "borrowings" between species, the viceroy butterfly, which is not itself noxious, exploits the monarch's defense since it looks like a monarch—a case of Batesian mimicry in which an edible species imitate an inedible one. In Mullerian mimicry, both or several inedible species actually imitate one another, widely broadcasting their defense by sharing common distinctive traits. **(Interactions: competition, predation and symbioses)**

56. a. iii, b. i, c. i, d. iii, e. ii, f. i, g. iii

    Commensalism refers to one partner in the relationship benefiting without harm or benefit to the other e.g. barnacles on whales. Mutualistic relationships (mutualism) benefit both members, like the bacteria normally found in the human gut or on human skin. Parasitism involves one organism becoming highly specialized and intimately dependent upon another for the completion of its life cycle such as tapeworms, or insect-borne diseases like malaria. Parasites vary greatly in their effect, and may do little or much harm to their hosts. Infections caused by viruses, bacteria and fungi are other examples of parasitism. **(Interactions: competition, predation and symbioses)**

57. Bird or insects have coevolved with the flowers they feed on and pollinate. Coevolution has actually been discussed above in all but name, as in instances of symbiosis. Coevolution is a concept that emphasizes the selective effects organism can have on each other in ecological relationships. Classic examples are pollinators like bees and flowers, or predators\herbivorves and prey. **(Coevolution)**

58. Carbon dioxide or sulfur from burning coal, pesticides washing off of farmland into a stream. Human activities can have detrimental effects on ecosystems. Especially with the advent of the industrial revolution, but actually beginning with concentrated human settlements, humans have released wastes and products into the environment that are either foreign in form or unusual in concentration or quantity. Mercury compounds, isolated from natural ores, are returned in toxic organic forms; carbon locked up for millennia is released in huge amounts in a short time period by burning fossil fuels. Such activities can

upset natural ecosystems by exceeding their capacity to process such materials and to regenerate. **(Pollution)**

59. Organisms—individuals or species—may interact in a number of different ways. When there is a common demand for a resource (food, nesting territory, reproductive mates), the result will be competition, especially if the resource is limited. Gause used this point to formulate his competitive exclusion principle: two species cannot coexist indefinitely in the same niche, that is, if each species has the same resource requirements as the other. Long-term survival demands each species finds it own niche. When species possess an overlapping niche, natural selection over time will tend to accentuate the differences between the species, eventually reducing the degree of overlap. **(Interactions: competition, predation and symbioses)**

Grade Yourself

Circle the question numbers that you had incorrect. Then indicate the number of questions you missed. If you answered more than three questions incorrectly, you will then have to focus on that topic. (If a topic has less than three questions and you had at least one wrong, we suggest you study that topic also. Read your textbook, a review book, or ask your teacher for help.)

Subject: Ecology, Cycles, and Populations.

Topic	Question Numbers	Number Incorrect
Life on earth	1, 2, 3, 4, 5, 6, 7	
Water: oceans and freshwater	8, 9, 10, 11, 12, 14	
Ecosystems	13, 15	
Biomes	16, 17, 18, 19, 20, 21, 22, 23	
Bio- and geochemical cycles	24, 25, 26, 27, 28, 29	
Energy and trophic levels	30, 31, 32, 33, 34, 35, 36	
Succession	37, 38, 39	
Populations and communities	40, 41	
Regulation of population size	42, 43, 44, 47	
Interactions: competition, predation and symbioses	45, 46, 50, 51, 52, 53, 54, 55, 56, 59	
Life history	48	
r- and K-selection	49	
Coevolution	57	
Pollution	58	